Atlas of Design

Vol. 1 »

editors
Timothy R. Wallace
Daniel P. Huffman

published by
nacis
the North American
Cartographic Information Society

Atlas of Design, Volume One

Revised Edition, Second Printing, July 2023

Edited by Timothy R. Wallace and Daniel P. Huffman

Published by the North American Cartographic Information Society, *nacis.org*

The type in this book has been set in Avenir, by Adrian Frutiger.

Printed in the United States of America by John Roberts, Inc., Minneapolis, MN.

ISBN 979-8-9866856-1-8

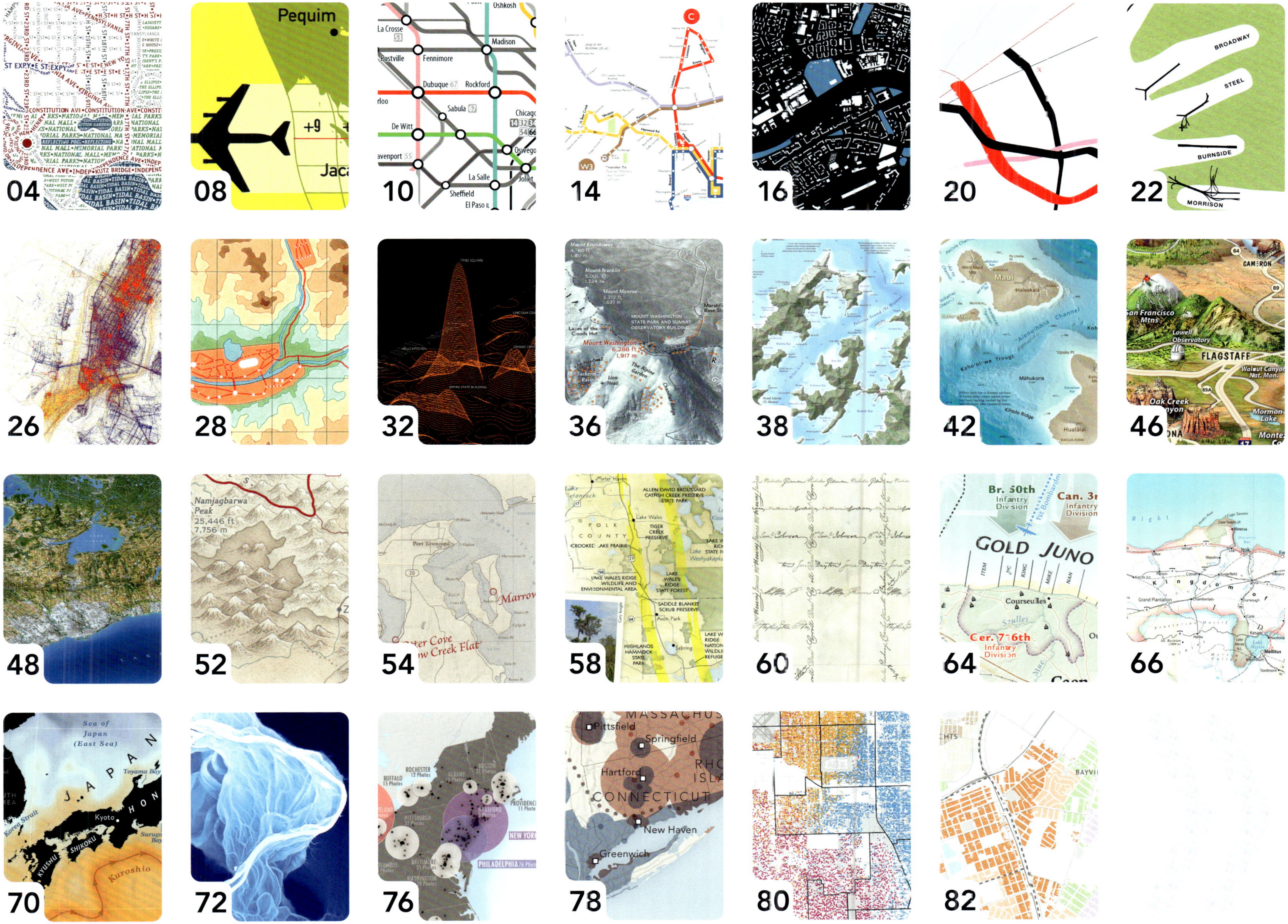
04
08
10
14
16
20
22
26
28
32
36
38
42
46
48
52
54
58
60
64
66
70
72
76
78
80
82

Preface

Nearly seven years ago, the North American Cartographic Information Society took a chance on this book. Publishing the first *Atlas of Design* was somewhat risky; it took a substantial portion of the organization's funds, and while we thought we had something special, it was by no means certain that people would want to buy it. By this time, attention (and money) had long been shifting toward eye-catching online interactive maps, and the convenience of reference maps that fit into your smartphone. Colleagues asked us why we didn't produce an eBook instead of a physical edition. But we believed in the unique experience of print. Our opening essay, reproduced on the following pages, is about how much aesthetics can matter in cartography. And one of the most important aesthetic decisions, beyond typefaces or line weights, is to choose to offer readers the chance to hold a piece of paper in their hands. We wanted the maps in this book to live in your home, not in the cloud. Instead of being briefly loaded into your phone and then dismissed to the ether, they exist in a purpose-built object that honors their quality.

It was daring to offer a book of printed maps in an era when print was in decline, but it seems that our optimism was warranted. As we celebrate the recent release of the fourth volume, and begin looking toward a fifth one, it's clear that the *Atlas* has found an audience, and is here to stay for the long term. It's been used in classrooms, spurred designers to try new ideas, and helped build connections in the cartographic world. It has been gratifying to see the project grow during each biennial cycle, with more submissions coming in from all over the planet, and more people eager to add it to their bookshelves. In the coming years, we hope it will continue to expand to become a global byword for inspirational cartography.

In your hands is a revised edition of the volume that started it all. It finally addresses a good-natured complaint we've long received: it now matches the physical dimensions of the subsequent volumes in the series, so you can shelve them all together without it looking odd. Thanks to the new dimensions, the book features an all-new layout, which better showcases the fine work of the cartographers that we honor within. And here it bears mentioning that none of this would have been possible if these people (and many more) hadn't made the effort to submit their work to a book no one had heard of. They, too, took a chance on the *Atlas*.

As did you, of course, by purchasing a copy—or perhaps more than one, over the years. Your support has helped us keep publishing these volumes, which gives us a chance to add to the conversation about what mapmaking is and what it can be. We're excited to see what the coming decades bring for the *Atlas of Design*.

Daniel P. Huffman & Timothy R. Wallace, 2019

Acknowledgements

We didn't know this going in, but it turns out that creating a book, even one where other people provide most of the content, is a pretty hefty undertaking. Bringing the *Atlas of Design* to life has taken the efforts of a great many people, and we want to acknowledge our debts.

First, we would like to thank the NACIS Board for their support—logistical, emotional, monetary, and otherwise—of this atlas. This book exists because of their vision and generosity.

Our call for participation generated an amazing response. We received over 140 works, from cartographers the world over. They provided us with the content needed to create a book, but more importantly, their enthusiasm for this project drove it forward. They showed us that it could work. Not all of them are featured here, but each of them helped make this project a reality through their willingness to contribute, and to get the word out to all corners of the globe. Thank you, each and every one of you.

This massive pile of submissions was reviewed by a panel of judges. They patiently read our lengthy instructions, recused themselves during conflicts of interest, and spent hours in the scrutiny of a great many maps. They made possible the painful task of paring our pile down into something that would fit into a book. Thank you Elbie Bentley, Tanya Buckingham, Hannah Fairfield, Bernhard Jenny, Nathaniel Kelso, and Tom Patterson.

For the actual assembly of the volume, we owe a tremendous debt of gratitude to the kind folks at the University of Wisconsin Cartography Lab. Their willingness to offer feedback and act as a constant resource for all things design made this book what it is.

Finally, this book is about the art of making beautiful maps, but its point would be lost without the words we use to describe them. Thanks to Kelly Loughman for her editorial prowess in helping make the words we use to describe our maps as powerful as the maps themselves.

'Why should we insist on beautiful maps?'

An Argument for Beauty

Timothy R. Wallace, Daniel P. Huffman

For many people, maps are solely about data. As far as these users are concerned, as long as the content is present and legible, the map "works." While they are not necessarily wrong in their assessment, a case can and should be made for the importance of aesthetics in cartography. Content must take a form, and *how* something is said is just as critical as *what* is said. Aesthetics are not simply the wrapping paper we put around the data when making a map. They are the framework in which that data is presented.

Historically, that framework has been varied, depending on the cartographer and the nature of the map. That is why, when we think generically of a "map," the term conjures up a number of disparate images. For many people—even modern cartographers who focus on things like interaction design and rich data visualization—one of the first images that the term brings to mind is an institutional graphic, such as a map issued by the USGS or the Ordnance Survey. These maps are highly structured in their creation, built first and foremost with strict adherence to thousands of cartographic conventions. Roads must be a certain width; coastlines a certain color; towns, hamlets, villages, and urban areas symbolized to precise specifications. Such maps may sometimes be attractive, but creative expression is subordinated to standardized specifications.

Then, there are the maps created with a personal touch, with painstakingly crafted aesthetics that work in concert with their purposes. These are the maps that, from the earliest days of cartography, have required the mapmaker to make countless artistic decisions throughout the process, and to strike a balance between artistry and pragmatism. For centuries, every map element was an opportunity for artistic expression—a chance to give a novel aesthetic to the data.

Nowadays, hardly any maps are left untouched by algorithms and automation. Reference maps on the Web and navigational maps designed for mobile devices are obvious examples, but they are not alone. Whether it is with a software package, a code library or a cloud application, more and more maps are being created with a digital stencil, leaving the business of look and feel to someone or some*thing* else.

At the same time, technology does have a plus side. Many—if not all—of the maps featured in this volume benefitted from such conveniences. But as the maps presented here prove, it is with the help of the cartographer's creative sensibilities that algorithms and automation can contribute to brilliant representations of space.

Design and aesthetics matter, because form is not secondary to function; form is integral to function. A map cannot function if it remains unread. To truly engage map users requires that we present them with something worth looking at. Something that they will want to spend time studying. Something that acknowledges the human need for beauty. Something that causes the user to think about the map in terms beyond whether or not it simply "works."

Ignoring aesthetics means ignoring the senses of the reader, dehumanizing the final product. It yields maps that are consulted only out of necessity and rarely shared. When the data is not pleasant to look at, it reaches far fewer eyes. If we want to understand why so many people have a poor knowledge of geography, we might look at the maps being used to educate them. These maps may be accurate, but are they attractive and engaging? We need maps that people enjoy spending time with; maps that they'll voluntarily look at long enough to learn something.

We cannot ignore aesthetics, then, if we want people to get the most out of our maps. But delving into aesthetics can make mapmakers uncomfortable, because everyone has different tastes. We may choose a style that attracts one reader while alienating another, and it may ultimately seem easier not to think about aesthetics at all. The neutrality offered by defaults or generic conventions feels safer. But it is better for a map to be appreciated by many and disliked by a few, than it is to be regarded as mundane by everyone. People gain little from unobtrusive maps, which sit idly in drawers or are briefly skimmed over in books. Meanwhile, the self-assured ones, which have no shame in admitting that they are a form of art, will be admired on museum walls or go halfway around the Internet in a day.

Of course, a successful cartographic creation cannot be judged solely on its visual appeal. If a map is intended to be a navigational aid, for instance, it must effectively convey appropriate directional information to succeed. If a map is made to indicate the distances between points of interest, it must use a certain projection to succeed. Still, while the success of a specific map may hinge on the use of the appropriate conventions, any map can be ruined by poor design.

The works in this collection are the counterargument to sterile, machine-driven, default-heavy, data-first mapmaking. They effectively impart their messages using clever, attractive designs that keep the viewer engaged, wanting to know more, to see more. They are in turn elegant, simple, bold, and complex, and demonstrate that there are myriad ways to make an artful map.

This is a book about the heart of cartography, and its future. In an era when mapmaking is being turned over to quick, cheap, insensate algorithms, this atlas reminds us of the power and beauty of well-executed cartographic design carried out by talented people. It is meant to honor those individuals featured, and to loudly proclaim that *what they are doing is important.*

Their efforts result in the kinds of maps that inspire people to action, to conversation, and to make more maps. They are the answer for anyone who asks, "What do cartographers do?"

We care about how the map looks.

Design and aesthetics matter, because form is not secondary to function; form is integral to function.

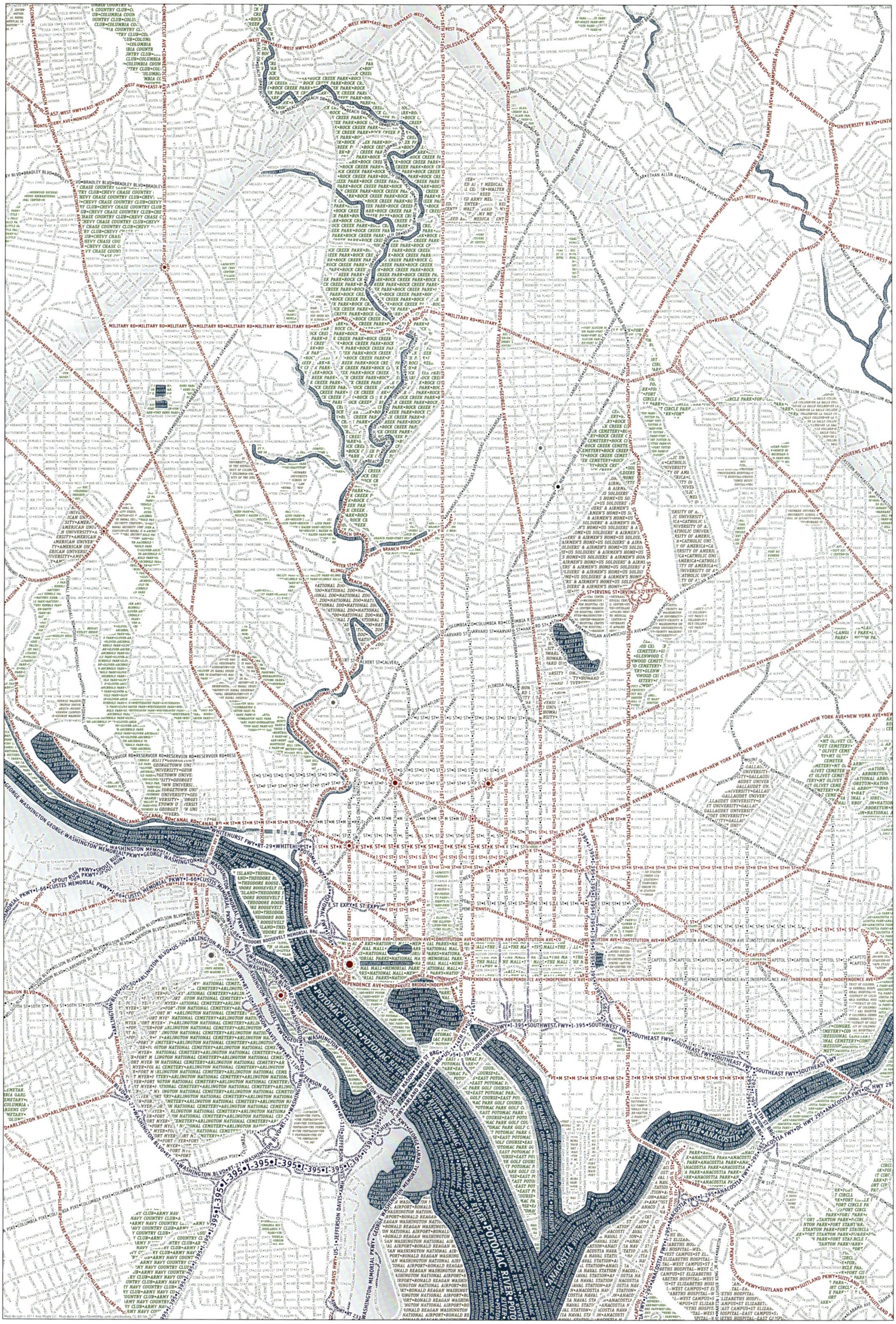

TON, D.C. • WASHINGTON, D.C. • WASHI

Geographic accuracy is paramount, but the subtle artistry of the maps lies in the elements that the geography comprises.

« Washington, D.C.

Ben Sheesley, David Heyman, Andy Woodruff, Mark Harrower — Axis Maps

Typography is an essential part of cartography. Words on maps provide reference, reveal data, and add meaning to graphical designs. Map typography requires great care; attention must be paid to hierarchy, legibility, placement, and aesthetics, which all impact the success and beauty of a map. We began designing purely typographic maps, in which physical features are symbolized by repeated text displaying their names, in part as an amusing departure from typical map design, but also as an embrace of the power of map typography.

Typography itself is also an art form, both in terms of the construction of letters and their arrangement on the page. Our maps are a combination of typographic art and reference map, in contrast to their more abstract peers in the world of typographic map art. A close-up look reveals the careful placement of words and choices of typefaces and styles, but at a few paces back, the many thousands of glyphs come together to form a simple city street map. In this way the design is intricate, but deliberately conservative. Geographic accuracy is paramount, but the subtle artistry of the maps lies in the elements that the geography comprises—perhaps in the use of varied type styles, in the way words intersect, or in occasional flair, like the flowing text of water bodies.

Type families are limited to two, consistent with some cartographic conventions. Streets use a sans-serif typeface, and a narrow style when possible to reduce white space within the text and improve overall contrast. For similar reasons, the use of capital letters throughout reduces white space and makes for even-looking lines and solid-looking fills. Streets fall into four categories based on their classification in OpenStreetMap data, and from these a hierarchy is achieved with bold styles and different sizes and colors. Water and other areal features are a serif typeface, frequently varying in size and orientation for water in order to create a fluid appearance.

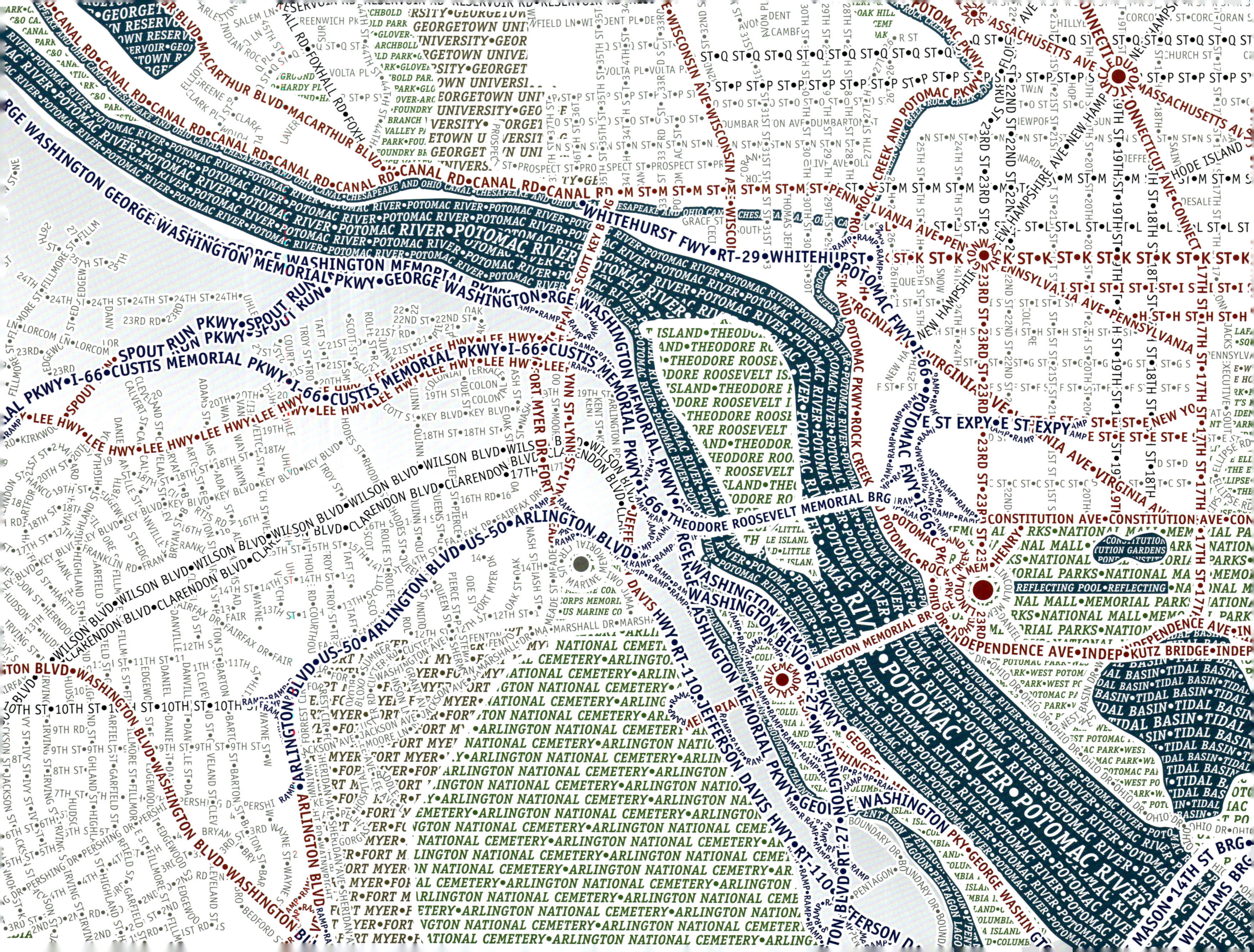

POTOMAC RIVER
GEORGETOWN RESERVOIR
CANAL RD
MACARTHUR BLVD
CHESAPEAKE AND OHIO CANAL
GEORGETOWN UNIVERSITY
WISCONSIN AVE
ROCK CREEK AND POTOMAC PKWY
MASSACHUSETTS AVE
CONNECTICUT AVE
NEW HAMPSHIRE AVE
PENNSYLVANIA AVE
WHITEHURST FWY
RT-29
GEORGE WASHINGTON MEMORIAL PKWY
SPOUT RUN PKWY
CUSTIS MEMORIAL PKWY
I-66
LEE HWY
WILSON BLVD
CLARENDON BLVD
ARLINGTON BLVD
US-50
THEODORE ROOSEVELT ISLAND
THEODORE ROOSEVELT MEMORIAL BRG
POTOMAC FWY
VIRGINIA AVE
E ST EXPY
CONSTITUTION AVE
NATIONAL MALL
MEMORIAL PARKS
CONSTITUTION GARDENS
REFLECTING POOL
LINCOLN MEM
INDEPENDENCE AVE
KUTZ BRIDGE
TIDAL BASIN
WEST POTOMAC PARK
ARLINGTON MEMORIAL BRG
ARLINGTON NATIONAL CEMETERY
FORT MYER
FORT MYER DR
JEFFERSON DAVIS HWY
RT-110
RT-27
WASHINGTON BLVD
MARINE CORPS MEMORIAL
PENTAGON
BOUNDARY DR
14TH ST BRG
MASON
WILLIAMS BRG

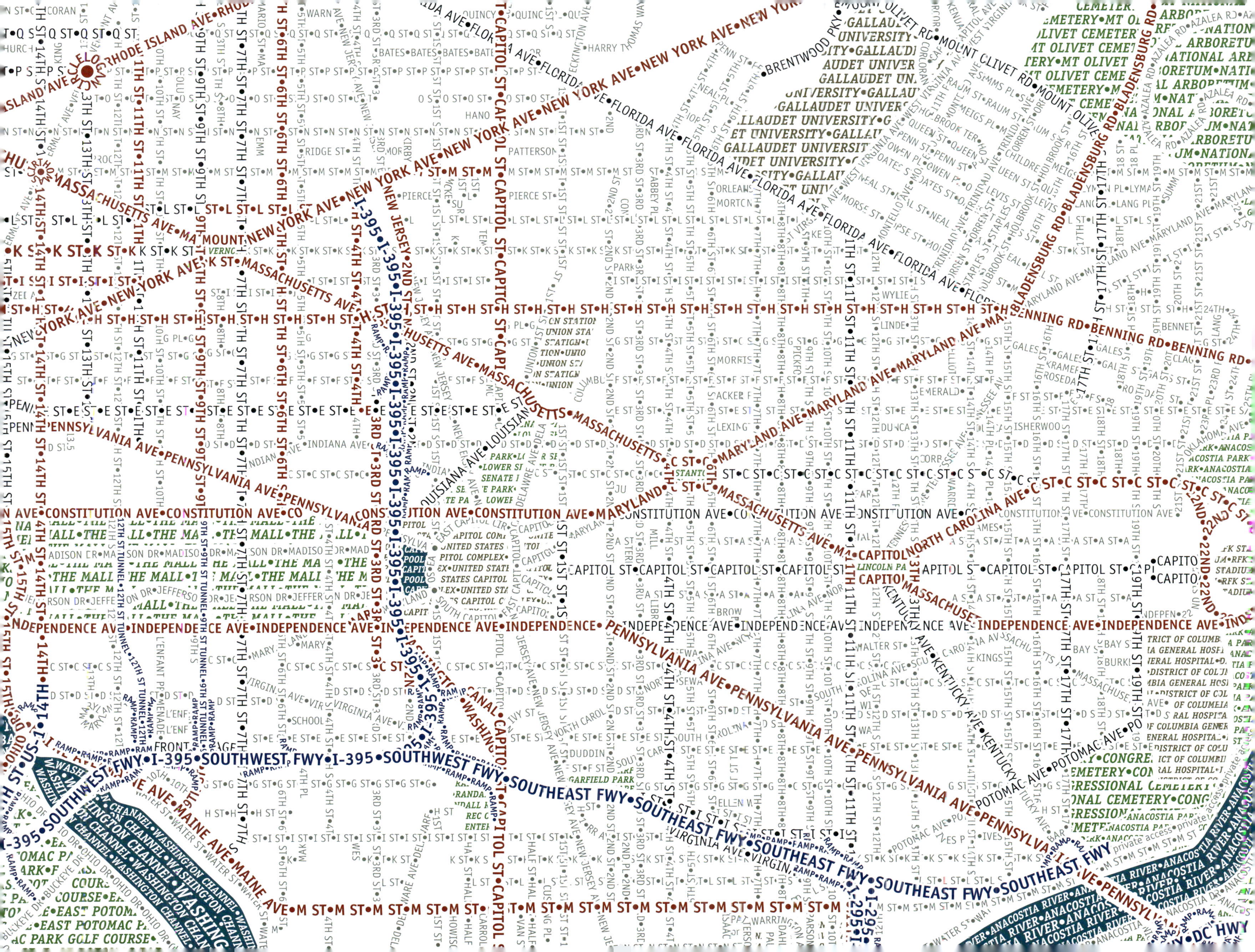

RHODE ISLAND AVE
LOGAN CIRCLE
NEW YORK AVE
FLORIDA AVE
MOUNT OLIVET RD
BRENTWOOD PKY
GALLAUDET UNIVERSITY
MT OLIVET CEMETERY
NATIONAL ARBORETUM
BLADENSBURG RD
MASSACHUSETTS AVE
MOUNT VERNON
K ST
I ST
H ST
I-395
NEW JERSEY AVE
BENNING RD
MARYLAND AVE
UNION STATION
PENNSYLVANIA AVE
E ST
INDIANA AVE
LOUISIANA AVE
C ST
CONSTITUTION AVE
THE MALL
MADISON DR
JEFFERSON DR
UNITED STATES CAPITOL
CAPITOL COMPLEX
CAPITOL ST
NORTH CAROLINA AVE
LINCOLN PARK
INDEPENDENCE AVE
KENTUCKY AVE
DISTRICT OF COLUMBIA GENERAL HOSPITAL
SOUTHWEST FWY
SOUTHEAST FWY
POTOMAC AVE
CONGRESSIONAL CEMETERY
MAINE AVE
WASHINGTON CHANNEL
WATER ST
OHIO DR
EAST POTOMAC PARK GOLF COURSE
VIRGINIA AVE
M ST
ANACOSTIA RIVER
ANACOSTIA PARK
DC HWY
I-295

Hora Mundial »

Eduardo Asta, Vincenzo Scarpellini

» *We also wanted to create a time zone map that . . . would recall a time when flying was a luxury—a herald of a modern way of life.*

When Vincenzo Scarpellini asked me to help him with an annual report he was doing for Infraero, the Brazilian airport authority, I created the infographics, including this time zone map. It was originally designed to be inserted in a Moleskine-size agenda, but eventually was only published inside the report.

Several factors influenced the way we constructed our map. Since many maps are Europe-centered, and our readers were Brazilians, we wanted a map with Brazil in its center. We also wanted to create a time zone map that is easy to understand, and that would recall a time when flying was a luxury—a herald of a modern way of life.

An Erik Nitsche poster for General Dynamics inspired us, too. It featured a Goode Homolosine world map with two airplanes crossing it in orthogonal directions, just above Brazil. That map impressed us. It was beautiful, unusual, rich, and very modern—and it had the meridians we needed. Its airplane running north-south did not work for us, so we replaced it with two west-east airplanes, each traveling in the opposite direction. This change took us one step closer to making our map easier to read.

Adapting the J. Paul Goode projection so that Brazil was in the center was a complicated but necessary step. Why? Because every one of us starts reading a map from our main point of reference. Someone from the United States starts reading a time zone map from the United States. Someone from Japan starts reading from Japan. Our putting Brazil in the center allowed readers to add and subtract hours in the simplest way.

In the end, we did not use Nitsche's colors. We decided on yellow and red (although there is another version of this map with a deep blue background, and orange strokes and fills in front). We refined our drawing too, so it looks almost technical, and the rulers were helpful in giving the map a precise, modernist look. Finally, we recovered Erik's vertical strip and used it to symbolize Brazil.

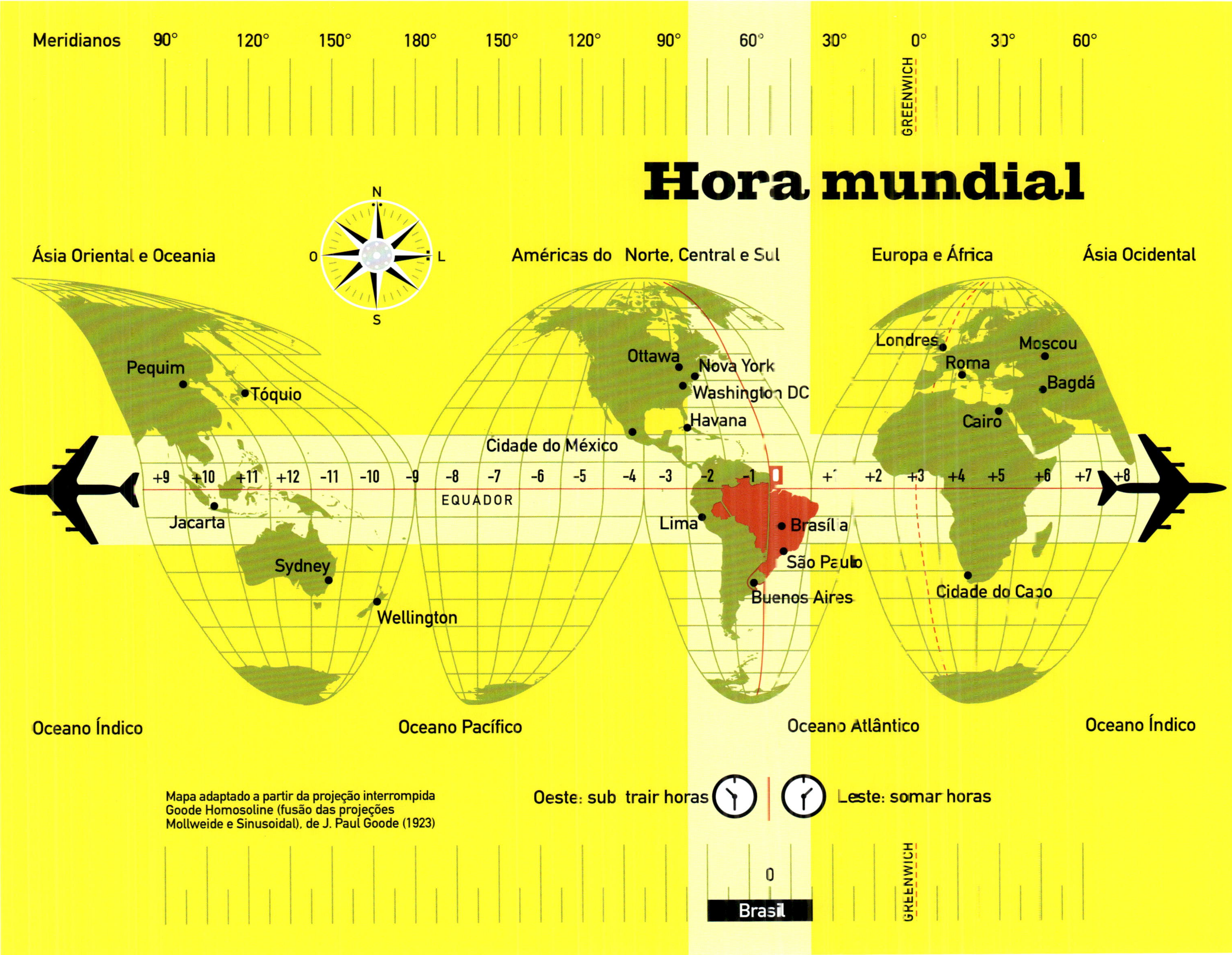
Hora mundial
Meridianos
90° 120° 150° 180° 150° 120° 90° 60° 30° 0° 30° 60°
GREENWICH
Ásia Oriental e Oceania
Américas do Norte, Central e Sul
Europa e África
Ásia Ocidental
N
S
O
L
Pequim
Tóquio
Jacarta
Sydney
Wellington
Cidade do México
Ottawa
Nova York
Washington DC
Havana
Lima
Brasília
São Paulo
Buenos Aires
Londres
Roma
Moscou
Bagdá
Cairo
Cidade do Cabo
+9 +10 +11 +12 -11 -10 -9 -8 -7 -6 -5 -4 -3 -2 -1 0 +1 +2 +3 +4 +5 +6 +7 +8
EQUADOR
Oceano Índico
Oceano Pacífico
Oceano Atlântico
Oceano Índico
Mapa adaptado a partir da projeção interrompida Goode Homosoline (fusão das projeções Mollweide e Sinusoidal), de J. Paul Goode (1923)
Oeste: subtrair horas
Leste: somar horas
0
Brasil
GREENWICH

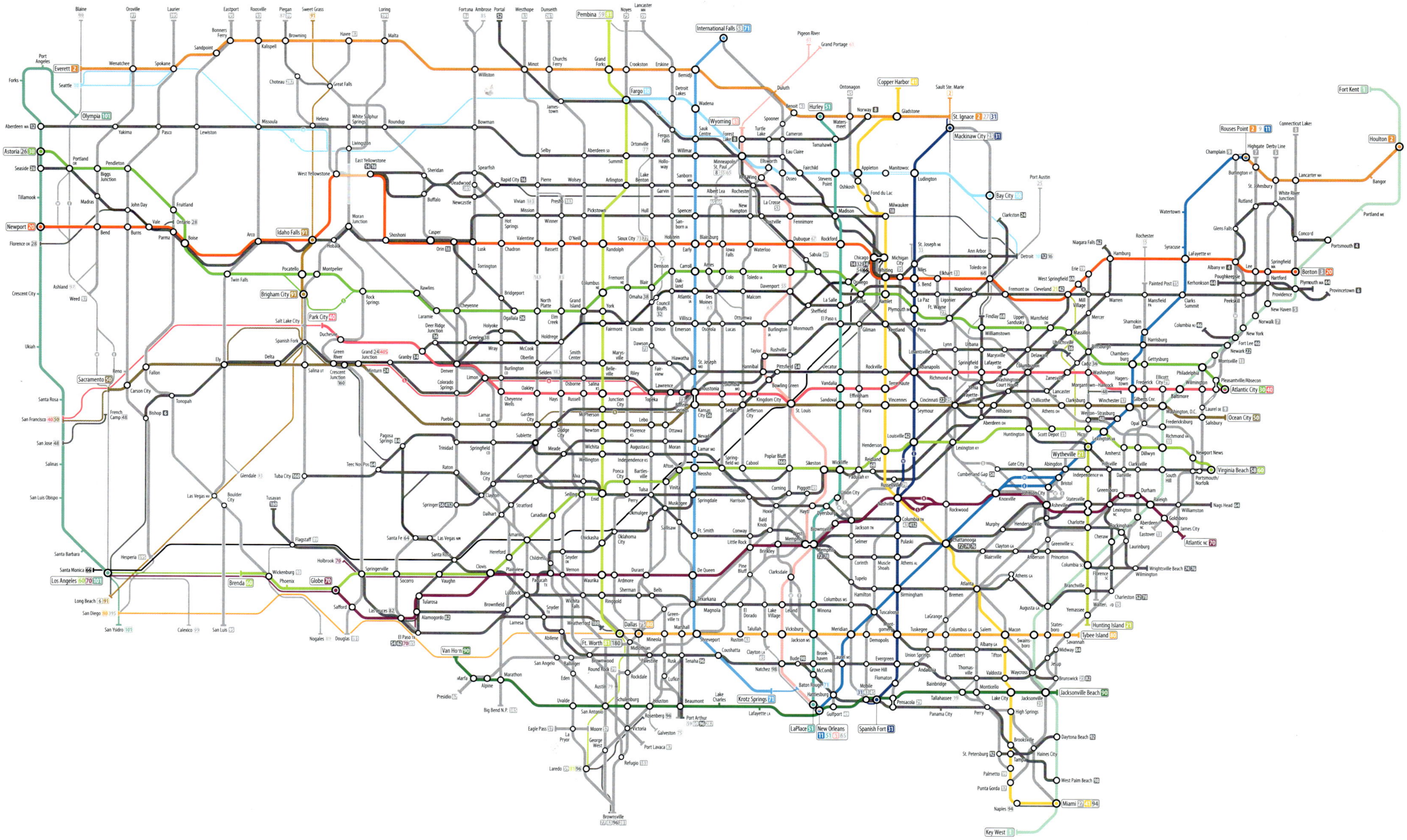

U.S. ROUTES

- Current route
- Historical route
- Routes not signed through Yellowstone National Park: implied route shown
- Ferry route

1 Fort Kent ME – Key West FL
2 **West:** Everett WA – St. Ignace MI
2 **East:** Rouses Point NY – Houlton ME
3 Connecticut Lakes NH – Boston MA
4 Albany NY – Portsmouth NH
5 Derby Line VT – New Haven CT
6 Bishop CA – Provincetown MA
7 Highgate Springs VT – Norwalk CT
8 Forest Lake MN – Norway MI
9 Champlain NY – Laurel DE
10 Fargo ND – Bay City MI

11 Rouses Point NY – New Orleans LA
12 Aberdeen WA – Detroit MI
13 Morrisville PA – Eastover NC
14 East Yellowstone WY – Chicago IL
15 Painted Post NY – Walterboro SC
16 East Yellowstone WY – Rapid City SD
17 Winchester VA – Punta Gorda FL
18 Orin WY – Milwaukee WI
19 Erie PA – Palmetto FL
20 Newport OR – Boston MA

21 Wytheville VA – Hunting Island SC
22 Cincinnati OH – Newark NJ
23 Mackinaw City MI – Jacksonville FL
24 Minturn CO – Clarkston MI
25 Cincinnati OH – Brunswick GA
26 Seaside OR – Ogallala NE
27 Fort Wayne IN – Miami FL
28 Florence OR – Ontario OR
29 Ellicott City MD – Pensacola FL
30 Astoria OR – Atlantic City NJ

31 Mackinaw City MI – Spanish Fort AL
32 Council Bluffs IA – Chicago IL
33 Elkhart IN – Richmond VA
34 Granby CO – Chicago IL
35 Michigan City IN – Scott Depot WV
36 Deer Ridge Junction CO – Uhrichsville OH
37
38 Greeley CO – Omaha NE
39
40 Park City UT – Atlantic City NJ

41 Copper Harbor MI – Miami FL
42 Louisville KY – Cleveland OH
43 Columbia TN – Mobile AL
44 Kerhonkson NY – Plymouth MA
45 Ontonagon MI – Mobile AL
46 Columbia NJ – Fort Lee NJ
47
48 Weston WV – Strasburg VA
48 **1926–1931:** San Jose CA – French Camp CA
48 **1973–1981:** Morgantown WV – Hancock VA
49 Piggott AR – Gulfport MS
50 Sacramento CA – Ocean City MD

51 Hurley WI – LaPlace LA
52 Portal ND – Charleston SC
53 International Falls MN – La Crosse WI
54 El Paso TX – Pittsfield IL
55 Minneapolis MN – Davenport IA
56 Springer NM – Kansas City MO
57 Moore TX – Eagle Pass TX
58 Cumberland Gap TN – Virginia Beach VA
59 Lancaster MN – Laredo TX
60 Brenda AZ – Virginia Beach VA

61 Wyoming MN – New Orleans LA
62 El Paso TX – Niagara Falls NY
63 Benoit WI – Ruston LA
64 Teec Nos Pos AZ – Nags Head NC
65 Albert Lea MN – Clayton LA
66 Santa Monica CA – Chicago IL
67 Sabula IA – Presidio TX
68 Reidland KY – Findlay OH
69 Albert Lea MN – Port Arthur TX
70 Globe AZ – Atlantic NC

71 International Falls MN – Krotz Springs LA
72 Memphis TN – Chattanooga TN
73 Dawson NE – Bonner Springs KS
74 Chattanooga TN – Wrightsville Beach NC
75 Noyes MN – Dallas TX
76 Chattanooga TN – Wrightsville Beach NC
77 Sioux City IA – Brownsville TX
78 Memphis TN – Charleston SC
79 Russellville KY – Round Rock TX
80 Dallas TX – Tybee Island GA

81 Pembina ND – Fort Worth TX
82 Alamogordo NM – Brunswick GA
83 Westhope ND – Brownsville TX
84 Pagosa Springs CO – Midway GA
85 Fortuna ND – El Paso TX
86
87 Havre MT – Port Lavaca TX
88
89 Piegan MT – Flagstaff AZ
90 Van Horn TX – Jacksonville Beach FL

91 Idaho Falls ID – Brigham City UT
92 St. Petersburg FL – Daytona Beach FL
93 Roosville MT – Wickenburg AZ
94 Naples FL – Miami FL
95 Eastport ID – San Luis AZ
96 Tenaha TX – Port Arthur TX
96 **1926–1934:** Brownsville TX – Rosenburg TX
96 **1934–1939:** Laredo TX – Rosenburg TX
97 Oroville WA – Weed CA
98 Bude MS – West Palm Beach FL
99 Blaine WA – Calexico CA
100

101 Olympia WA – Los Angeles CA
160 Tuba City AZ – Poplar Bluff MO
180 Tusayan AZ – Weatherford TX
183 Presho SD – Refugio TX
191 Loring MT – Douglas AZ
281 Dunseith ND – Brownsville TX
287 Choteau MT – Port Arthur TX
385 Deadwood SD – Big Bend National Park TX
395 Laurier WA – Hesperia CA
412 Springer NM – Columbia TN

This map is the most complex that I have ever attempted, and represents well over 400 hours of research and design, spread over almost a year.

« U.S. Numbered Highways as a Subway Map

Cameron Booth

This map is the latest in a series that concentrates on reducing complex networks down to their most basic topological connections, and presents them in the familiar form of a subway map. This map focuses on the U.S. Highway system, the secondary network of major roads (after the newer Interstate Highways) that span the United States. In addition to extant roads, the map also uses thinner lines to show historic routes, such as the famous Route 66.

Visually, the map pays homage to the iconic London Underground Tube map, the forefather of the hundreds of diagrammatic maps used by transit systems around the world. I personally regard the Tube map as one of the greatest works of informational design ever, and one that has stood the test of time from the 1930s through to the modern day. Modeling my work on its aesthetics instead of another famous subway map, such as the Washington, D.C. Metro map, was an easy choice.

Almost every aspect of this map is influenced by the Tube map, right down to the logo at the bottom-left that playfully echoes the famous London Underground roundel. The only major difference lies in the typeface used; due to the complexity of my map, I needed a much narrower typeface than the Tube map's distinctive Johnston Sans.

This map is the most complex that I have ever attempted, and represents well over 400 hours of research and design, spread over almost a year. It probably represents the absolute limit of what this type of map can depict while still being both aesthetically appealing and information-rich.

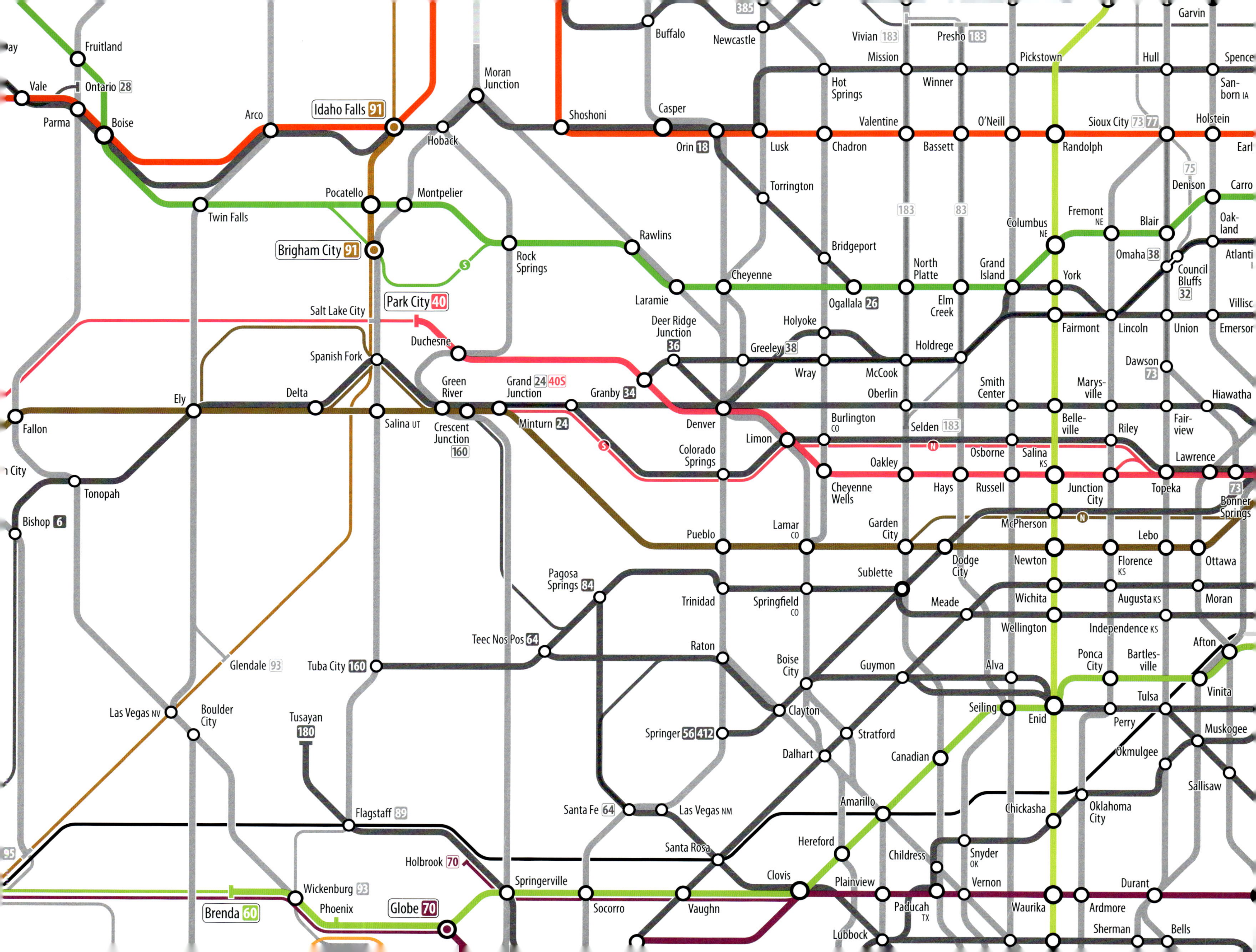

Fruitland
Vale
Ontario 28
Parma
Boise
Arco
Idaho Falls 91
Hoback
Moran Junction
Shoshoni
Buffalo
385
Newcastle
Casper
Orin 18
Lusk
Vivian 183
Presho 183
Mission
Hot Springs
Winner
Pickstown
Garvin
Hull
Spence
San-born IA
Valentine
Chadron
Bassett
O'Neill
Sioux City 73 77
Randolph
Holstein
Earl
Twin Falls
Pocatello
Montpelier
Brigham City 91
Rock Springs
Rawlins
Torrington
Bridgeport
183
83
75
Denison
Carro
Columbus NE
Fremont NE
Blair
Oak-land
Omaha 38
Atlanti
Council Bluffs 32
Cheyenne
Laramie
North Platte
Grand Island
York
Ogallala 26
Elm Creek
Park City 40
Salt Lake City
Duchesne
Deer Ridge Junction 36
Holyoke
Greeley 38
Holdrege
Fairmont
Lincoln
Union
Emerson
Villisc
Spanish Fork
Wray
McCook
Dawson 73
Green River
Grand 24 40S Junction
Granby 34
Smith Center
Marys-ville
Hiawatha
Ely
Delta
Oberlin
Fallon
Salina UT
Crescent Junction 160
Minturn 24
Denver
Burlington CO
Selden 183
Belle-ville
Riley
Fair-view
Limon
Osborne
Salina KS
Lawrence
City
Colorado Springs
Oakley
Tonopah
Cheyenne Wells
Hays
Russell
Junction City
Topeka
Bonner Springs
73
Bishop 6
McPherson
Lamar CO
Garden City
Lebo
Pueblo
Florence KS
Ottawa
Dodge City
Newton
Sublette
Pagosa Springs 84
Trinidad
Springfield CO
Meade
Wichita
Augusta KS
Moran
Wellington
Independence KS
Afton
Teec Nos Pos 64
Raton
Boise City
Ponca City
Bartles-ville
Glendale 93
Tuba City 160
Guymon
Alva
Vinita
Tulsa
Seiling
Enid
Perry
Las Vegas NV
Boulder City
Tusayan 180
Clayton
Muskogee
Springer 56 412
Stratford
Okmulgee
Dalhart
Canadian
Sallisaw
Flagstaff 89
Santa Fe 64
Las Vegas NM
Amarillo
Chickasha
Oklahoma City
95
Santa Rosa
Hereford
Childress
Snyder OK
Holbrook 70
Clovis
Plainview
Vernon
Durant
Springerville
Wickenburg 93
Socorro
Vaughn
Paducah TX
Waurika
Ardmore
Phoenix
Globe 70
Brenda 60
Lubbock
Sherman
Bells

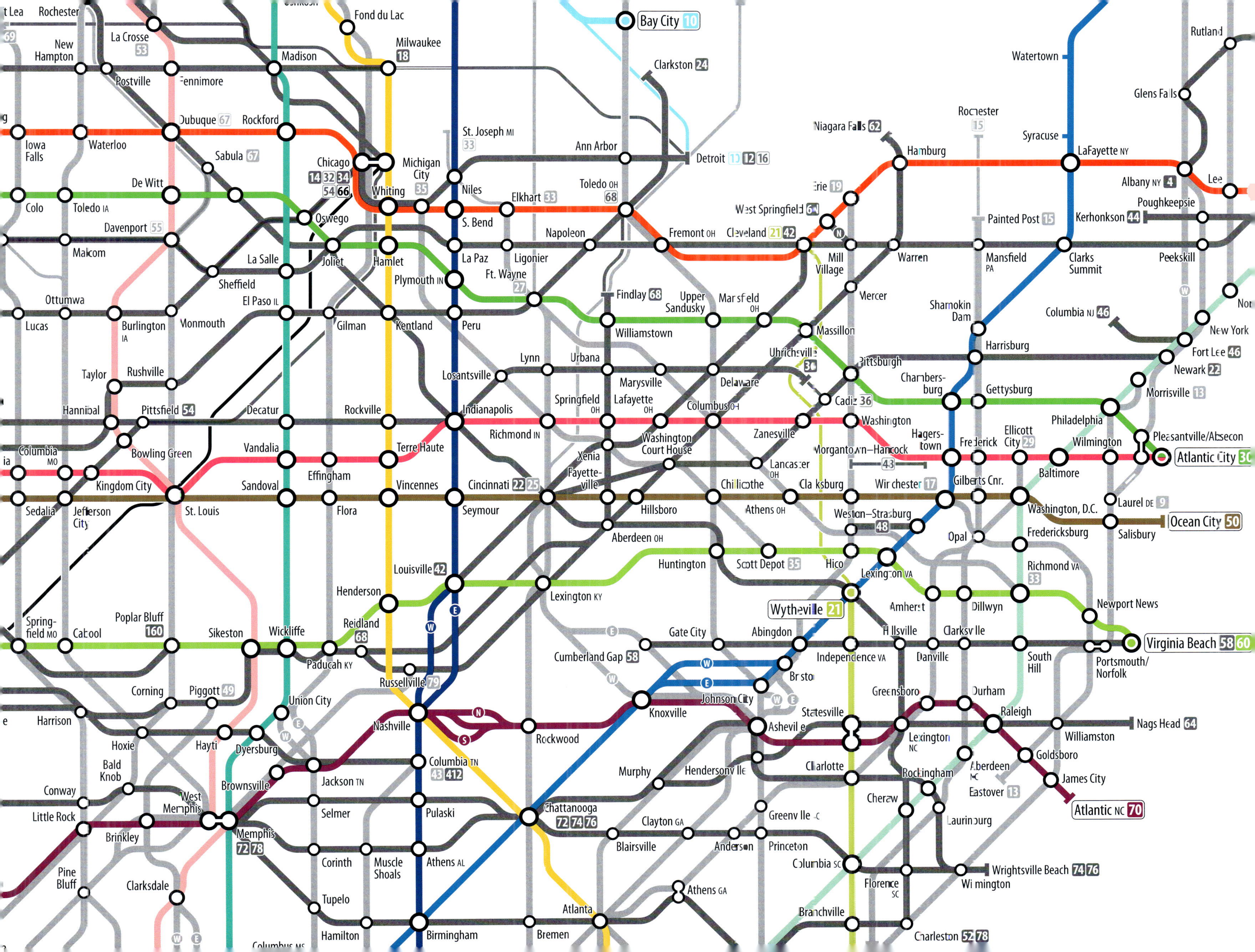

Rochester
La Crosse 53
Fond du Lac
Bay City 10
New Hampton
Postville
Fennimore
Madison
Milwaukee 18
Clarkston 24
Rutland
Watertown
Glens Falls
Dubuque 67
Rockford
Iowa Falls
Waterloo
Sabula 67
Chicago 14 32 34 54 66
Michigan City 35
Whiting
St. Joseph MI 33
Ann Arbor
Detroit 10 12 16
Niagara Falls 62
Hamburg
Rochester 15
Syracuse
LaFayette NY
Albany NY 4
Lee
De Witt
Colo
Toledo IA
Oswego
Niles
S. Bend
Elkhart 33
Toledo OH 68
Erie 19
West Springfield 61
Poughkeepsie
Kerhonkson 44
Painted Post 15
Davenport 55
Malcom
Joliet
Hamlet
La Salle
Sheffield
La Paz
Ligonier
Napoleon
Fremont OH
Cleveland 21 42
Mill Village
Warren
Mansfield PA
Clarks Summit
Peekskill
Plymouth IN
Ft. Wayne 27
Findlay 68
Upper Sandusky
Mansfield OH
Mercer
Shamokin Dam
Columbia NJ 46
Ottumwa
El Paso IL
Lucas
Burlington IA
Monmouth
Gilman
Kentland
Peru
Williamstown
Massillon
New York
Harrisburg
Fort Lee 46
Uhrichsville 36
Pittsburgh
Newark 22
Taylor
Rushville
Losantsville
Lynn
Urbana
Marysville
Delaware
Chambersburg
Gettysburg
Morrisville 13
Hannibal
Pittsfield 54
Decatur
Rockville
Indianapolis
Springfield OH
Lafayette OH
Columbus OH
Cadiz 36
Washington
Philadelphia
Pleasantville/Absecon
Richmond IN
Washington Court House
Zanesville
Hagerstown
Frederick
Ellicott City 29
Wilmington
Atlantic City 30
Columbia MO
Bowling Green
Vandalia
Terre Haute
Xenia
Fayetteville
Lancaster OH
Morgantown–Hancock 43
Baltimore
Kingdom City
Sandoval
Effingham
Vincennes
Cincinnati 22 25
Chillicothe
Clarksburg
Winchester 17
Gilberts Cnr.
Laurel DE 9
Sedalia
Jefferson City
St. Louis
Flora
Seymour
Hillsboro
Athens OH
Weston–Strasburg 48
Washington, D.C.
Ocean City 50
Opal
Salisbury
Aberdeen OH
Fredericksburg
Louisville 42
Huntington
Scott Depot 35
Hico
Lexington VA
Richmond VA 33
Henderson
Lexington KY
Wytheville 21
Amherst
Dillwyn
Newport News
Springfield MO
Poplar Bluff 160
Cabool
Sikeston
Wickliffe
Reidland 68
Paducah KY
Gate City
Abingdon
Hillsville
Clarksville
Cumberland Gap 58
Independence VA
Danville
South Hill
Virginia Beach 58 60
Portsmouth/Norfolk
Bristol
Russellville 79
Corning
Piggott 49
Union City
Johnson City
Greensboro
Durham
Harrison
Nashville
Knoxville
Statesville
Raleigh
Hoxie
Hayti
Dyersburg
Rockwood
Asheville
Lexington NC
Williamston
Nags Head 64
Bald Knob
Columbia TN 43 412
Charlotte
Aberdeen NC
Goldsboro
Conway
West Memphis
Brownsville
Jackson TN
Murphy
Hendersonville
Rockingham
Eastover 13
James City
Little Rock
Selmer
Pulaski
Chattanooga 72 74 76
Cheraw
Atlantic NC 70
Brinkley
Memphis 72 78
Clayton GA
Greenville SC
Laurinburg
Blairsville
Anderson
Princeton
Corinth
Muscle Shoals
Athens AL
Columbia SC
Wrightsville Beach 74 76
Pine Bluff
Clarksdale
Wilmington
Athens GA
Florence SC
Tupelo
Atlanta
Branchville
Hamilton
Birmingham
Bremen
Charleston 52 78
Columbus MS

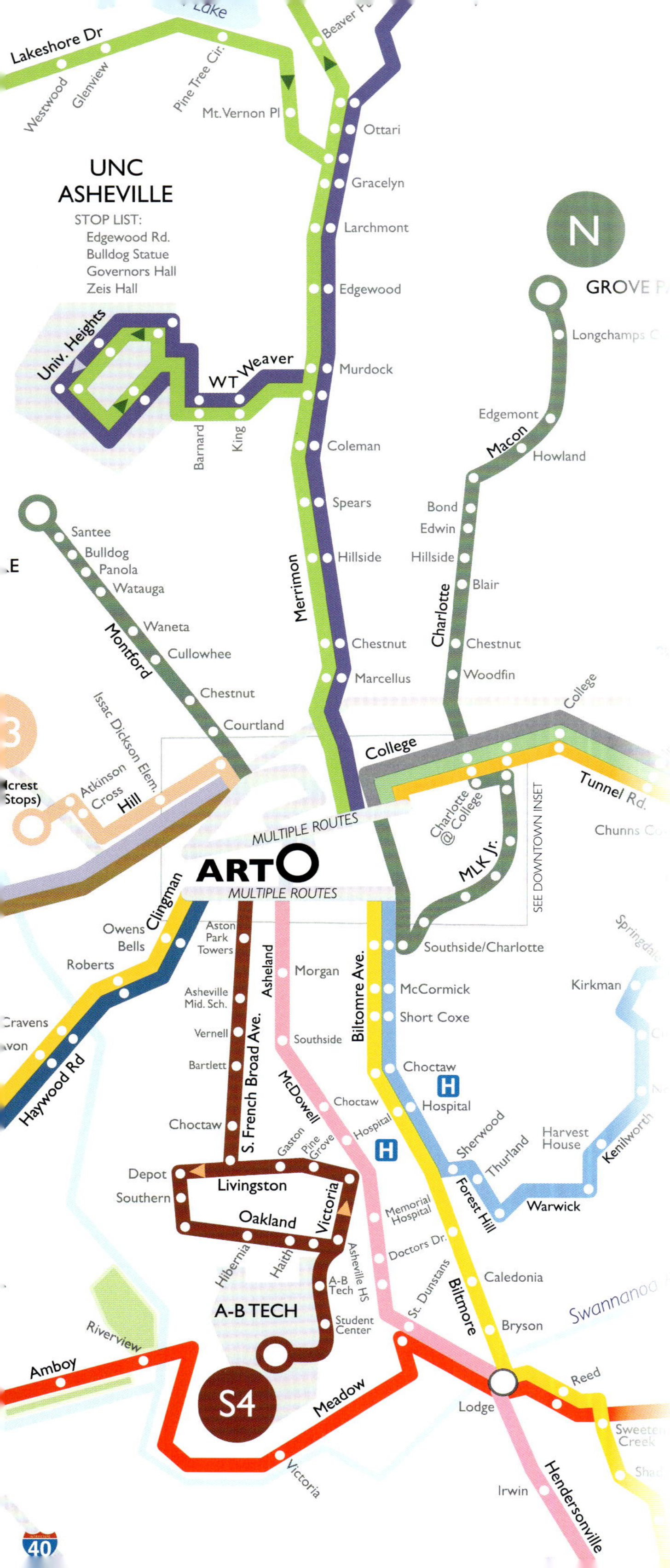

Asheville Redefines Transit: System Map »

Matt Forrest, Kate Chanba — Carticulate

The design parameters for this map were fairly simple: we wanted to embed a sense of place into the map, show all the stops clearly, and make it a high-level navigational tool. Asheville has a very unique identity, and our color and font choices for the map were influenced by a desire to have it fit in with the local scenery, as well as work alongside existing wayfinding tools in the city. Bringing this local identity into the map without sacrificing its navigational functionality was our biggest challenge, but an important one: as simple as fonts and colors are, they can speak volumes about a city.

The modern design school of transit maps was a major source of technical inspiration. Our symbols for the routes, stops, and bus lines were derived from timeless examples, and they work quite well for navigational purposes. This task is also helped by the significant role of landmarks in the map. As with many other transit maps, major features such as rivers and parks, as well as interstates and other highways, help to frame the network with an additional level of referential geography.

Representing the system clearly was a challenge, since Asheville has many distinct areas in which roads follow the mountainous geography of the city. We felt that because most users of the map would be familiar with the major corridors that the routes travel on, having every stop the buses make would be more helpful than having the complete street geography on the map. Although the system is small compared to other major American cities, users can see the full scope of the bus network in one map.

Our goals were cleanliness, functionality, and identity. In our opinion, these were all met to varying degrees in the final map. We don't think that any of them were completely met, but in combination they created a balanced mix.

ART Asheville Redefines Transit

CROSS TOWN	C	Louisiana - Haywood - State - Amboy - Biltmore Village - Fairview - Swannanoa Rd - Wood
	N	Klondyke - Montford - Downtown - MLK - Charlotte - Grove Park
NORTH	N1	ART Station - Merrimon - UNCA - Lakeshore
NORTH	N2	ART Station - Merrimon - UNCA - Beaverdam
NORTH	N3	ART Station - Chamber - Hillcrest
SOUTH	S1	ART Station - Biltmore- Hospital - Biltmore Village - London - Shiloh - Caribou - Rock Hill - Sweeten Creek
SOUTH	S2	ART Station - Biltmore - Hospital - Forest Hill - Kenilworth - Chunns Cove - Social Security
SOUTH	S3	ART Station - Asheland - McDowell - Biltmore Village - Hendersonville Rd. - Airport
SOUTH	S4	ART Station - S. French Broad - Depot - Livingston Heights - AB Tech
EAST	E1	ART Station - Tunnel - Asheville Mall - South Tunnel - Wallmart - Swannanoa River - VA
EAST	E2	ART Station - Asheville Mall - Haw Creek - Tunnel
WEST	W1	ART Station - Hilliard - Clingman - Haywood - PVA - Deaverview Area
WEST	W2	ART Station - Hilliard - Clingman - Haywood - PVA - Brevard - Biltmore Square Mall
WEST	W3	ART Station - Patton - Goodwill
WEST	W4	ART Station - Patton - New Leicester - Land of Sky
	170	ART Station - Tunnel Rd. - Warren Wilson - Swannanoa - Black Mountain

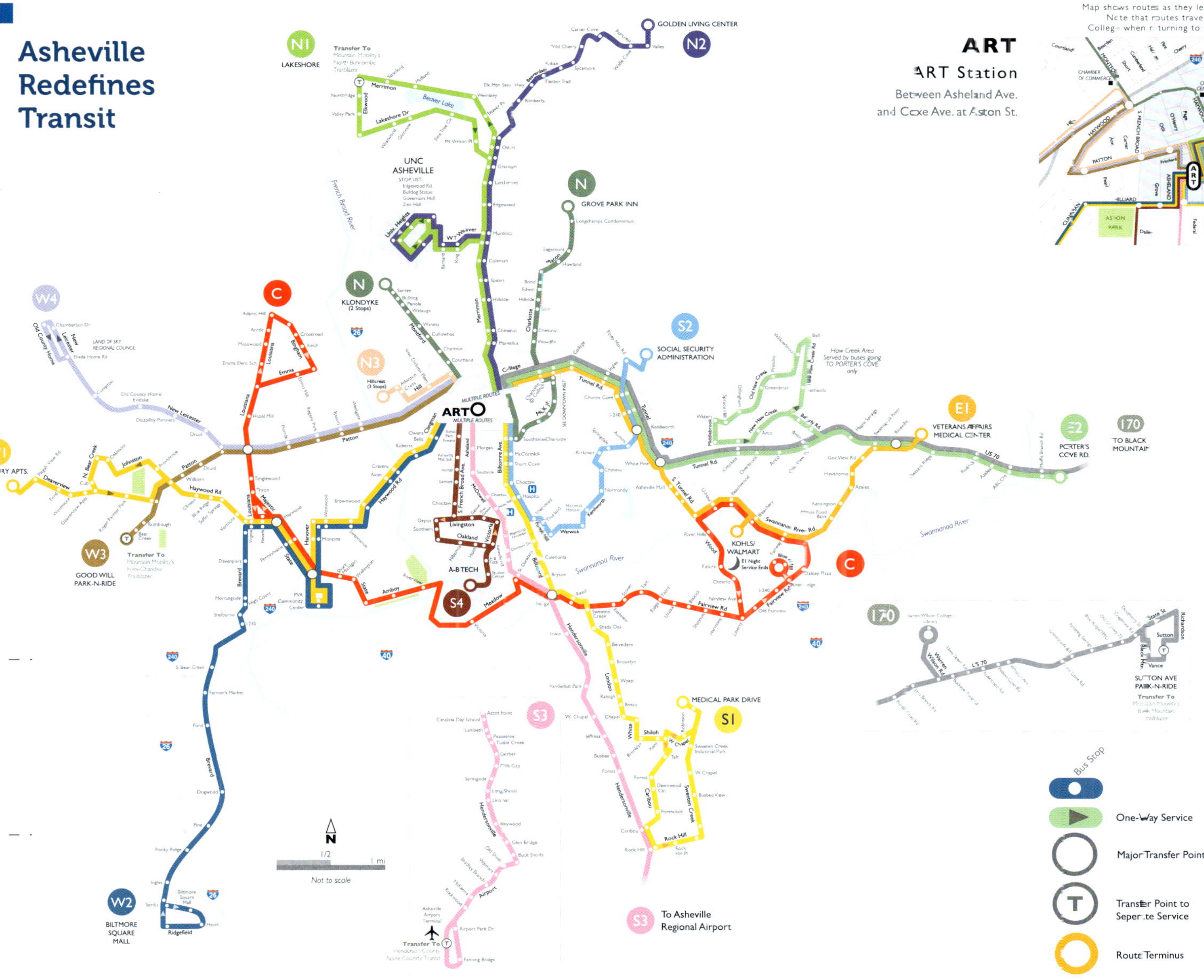

Downtown Asheville

Map shows routes as they leave outbound from the ART Station. Note that routes traveling using Patton outbound use College when returning to the ART Station on the return trip.

ART

ART Station

Between Asheland Ave. and Coxe Ave. at Aston St.

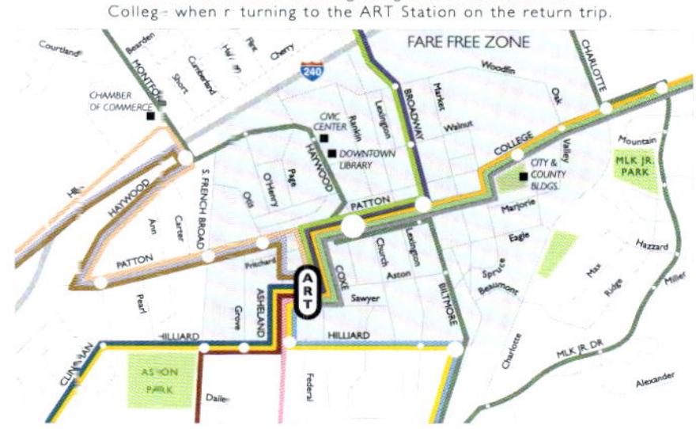

How to Ride the Bus

Boarding

Locate a bus stop sign along the route you are traveling on. Make sure you are on the side of the road in the direction you wish to travel. Have the exact fare ready. The bus will stop for you when you are standing by the stop.

If you plan on transferring to another route, ask the driver for a transfer ticket.

Departing

Touch the yellow "touch tape" on the bus before you arrive at your desired stop.

If you need wheelchair or other assistance, wait for the driver to assist you.

All buses have bike racks on the front of the bus. When departing, inform the driver you need to unload your bike.

Bus Stop

One-Way Service

Major Transfer Point

Transfer Point to Seperate Service

Route Terminus

Avon Haywood = Haywood Rd at Avon Ave

Note: Some stops are located at other nearby streets. Be sure to locate the bus stop sign and be on the side of the street in the direction you desire to travel.

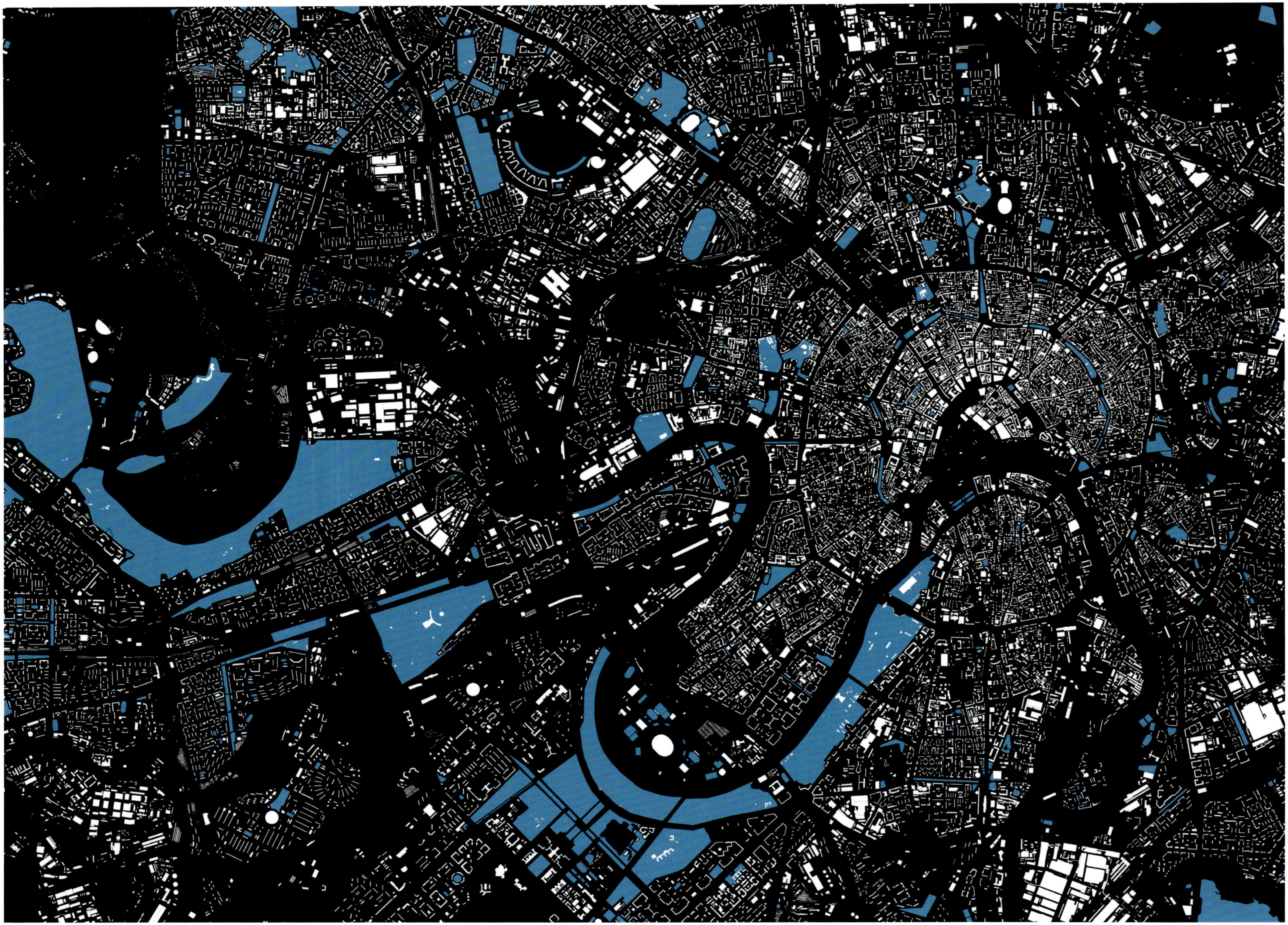

These are not maps for bombing your enemies, but then, maps are no longer a binary proposition, if they ever were.

« map=yes

Aaron Straup Cope — Stamen Design

map=yes is a love song to comic book artists, in particular Dave McKean, who authored and illustrated the graphic novel *Cages* in the 1990s. With few exceptions, the only colors in *Cages* are pale, slate blues that shadow some of the most beautiful black and white pen and ink line work I've ever seen. It is a wonderful example of the old adage about doing more with less.

map=yes began as a project that Stamen Design did for MapQuest in 2011 to highlight their newly released API services, which were part of their ongoing MapQuest Open and OpenStreetMap (OSM) initiatives. One of the APIs they released was a publicly accessible instance of the OSM Extended API, which is commonly referred to as "XAPI."

OSM consists of two databases that are derived from the sum of all the data collected by its members. The "rendering" database is a subset of the whole and used to produce the map that people see at openstreetmap.org. The "tagging" database is the unfiltered whole; every recorded bit of geographic and tag data in OSM.

XAPI provides programmatic access to the tagging database. This is useful because there's a treasure trove of data in those tags that would otherwise remain inaccessible to non-expert OSM users. Some of it is messy, some of it is silly, but the beauty of the tagging database is that same breadth of subject matter, and the multiplicity of its uses.

As a final deliverable, Stamen produced a set of open source tools for creating online ("slippy") maps, derived from one or more tag queries against the XAPI endpoint. To demonstrate the software, a series of maps was produced for 20 cities that consisted solely of features in OSM. They were tagged "building" (in white), "leisure" (in blue), or "trunk" (in gray).

These are not maps for bombing your enemies, but then, maps are no longer a binary proposition, if they ever were. We have spent the last 250 years chasing down the dragons that used to haunt the maps that came before. Since then we've uncovered the twin super-powers of community and networked computers. We will never run out of unknown territories to chart, so what does it mean to use these new tools to look for new dragons?

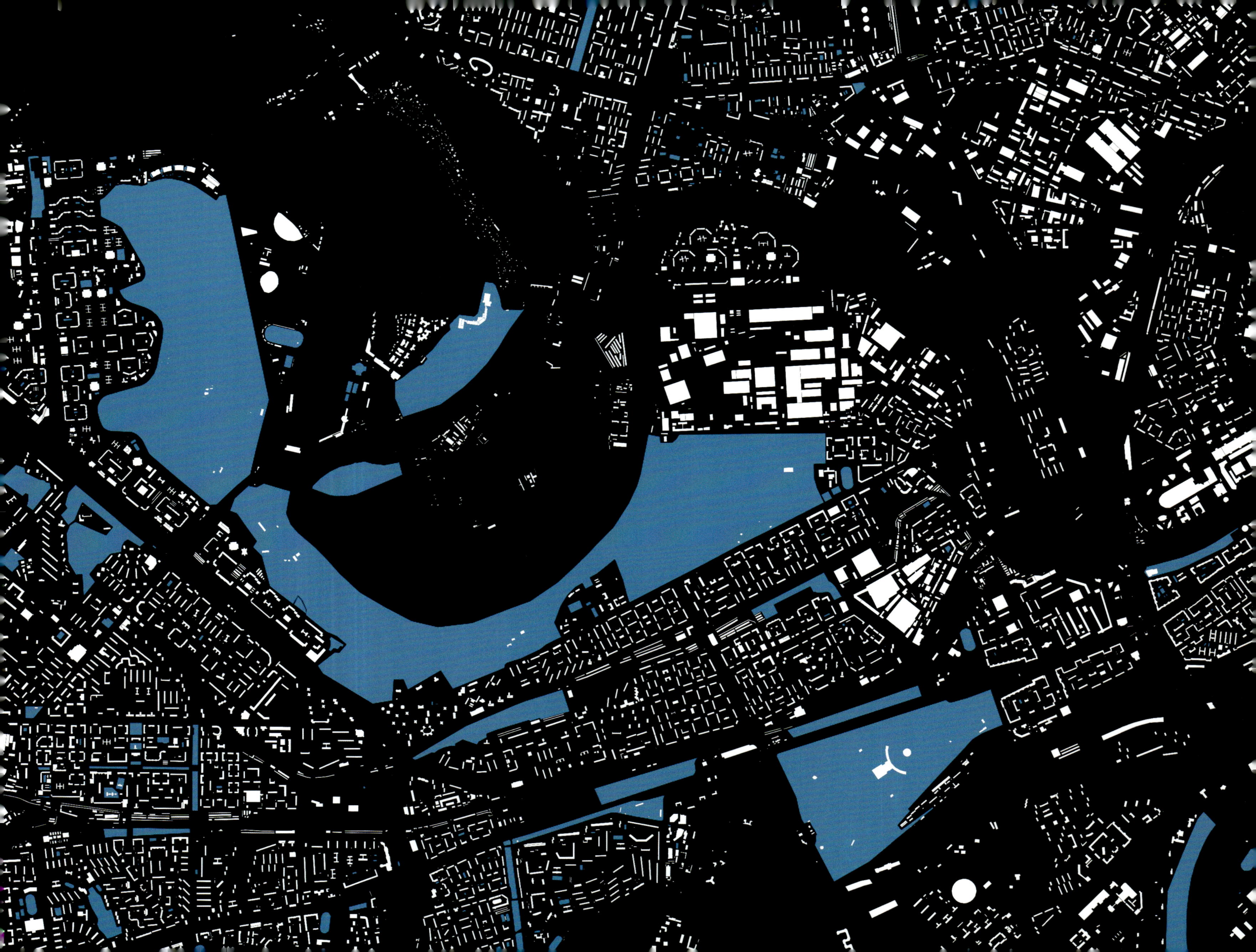

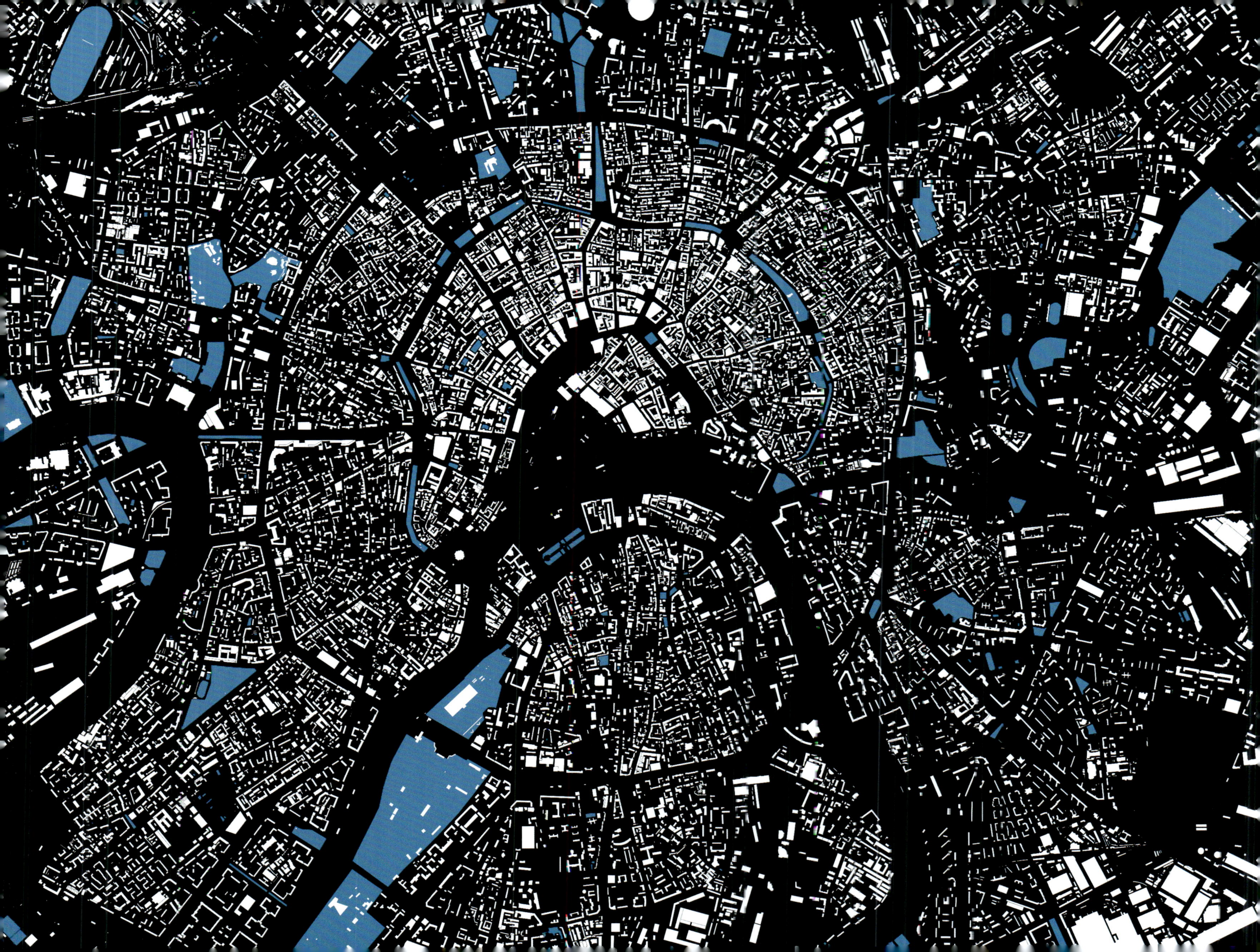

» *The aura of precision and comprehensiveness emanating from the familiar Michelin road map, which gives confidence to the traveler, is abandoned.*

figures INFRASTRUCTURES: An Atlas of Roads and Railways »

Bieke Cattoor, Bruno De Meulder

Excerpt from: Cattoor, Bieke and De Meulder, Bruno (2011). Figures Infrastructures. An Atlas of Roads and Railways. Amsterdam: SUN, 176p.

figures INFRASTRUCTURES is a re-cartography of sections E18–H16 of Michelin road map number 533, titled *Regional Benelux*. The content of this single map is recomposed into the format of a book, and thereby becomes an atlas. As the lines on the Michelin map are more or less straightforward representations of the existing transport infrastructures, they remain unaltered. They are, however, subject to a rearrangement: the given lines are forced into a new medium. By transferring them into the format of the atlas, the conception of the infrastructural networks takes on a renewed shape. The aura of precision and comprehensiveness emanating from the familiar Michelin road map, which gives confidence to the traveler, is abandoned. In the space of the book, stories are densely stacked, and the Cartesian projection moves over for multiple compositions.

figures INFRASTRUCTURES rearranges the infrastructural lines into what we could call a topological atlas structure, as opposed to the conventional atlas structure which can be considered topographical. This topological atlas structure highlights the dynamic interdependence of the infrastructural networks (roads and railroads) as they develop historically over time into complex interrelated systems.

Along the diagonal of the page, the re-cartography is confronted with images of historical maps and atlases. They are the references of the book. Not only did they provide the historical facts and data concerning the evolution of the infrastructures, they are the main suppliers of the cartographic knowledge that is absorbed in figures. This absorption of existing cartographic knowledge acquires a variety of modalities: from the transference of a cartographic decompositioning strategy to stylistic mimicry as a means of visual historical referencing. It is a mode of accumulating the knowledge of the larger cartographic community within a new atlas project.

figures INFRASTRUCTURES reconstructs the view, showing the potential of the cartographic atlas to reinterpret the everyday urbanized territory.

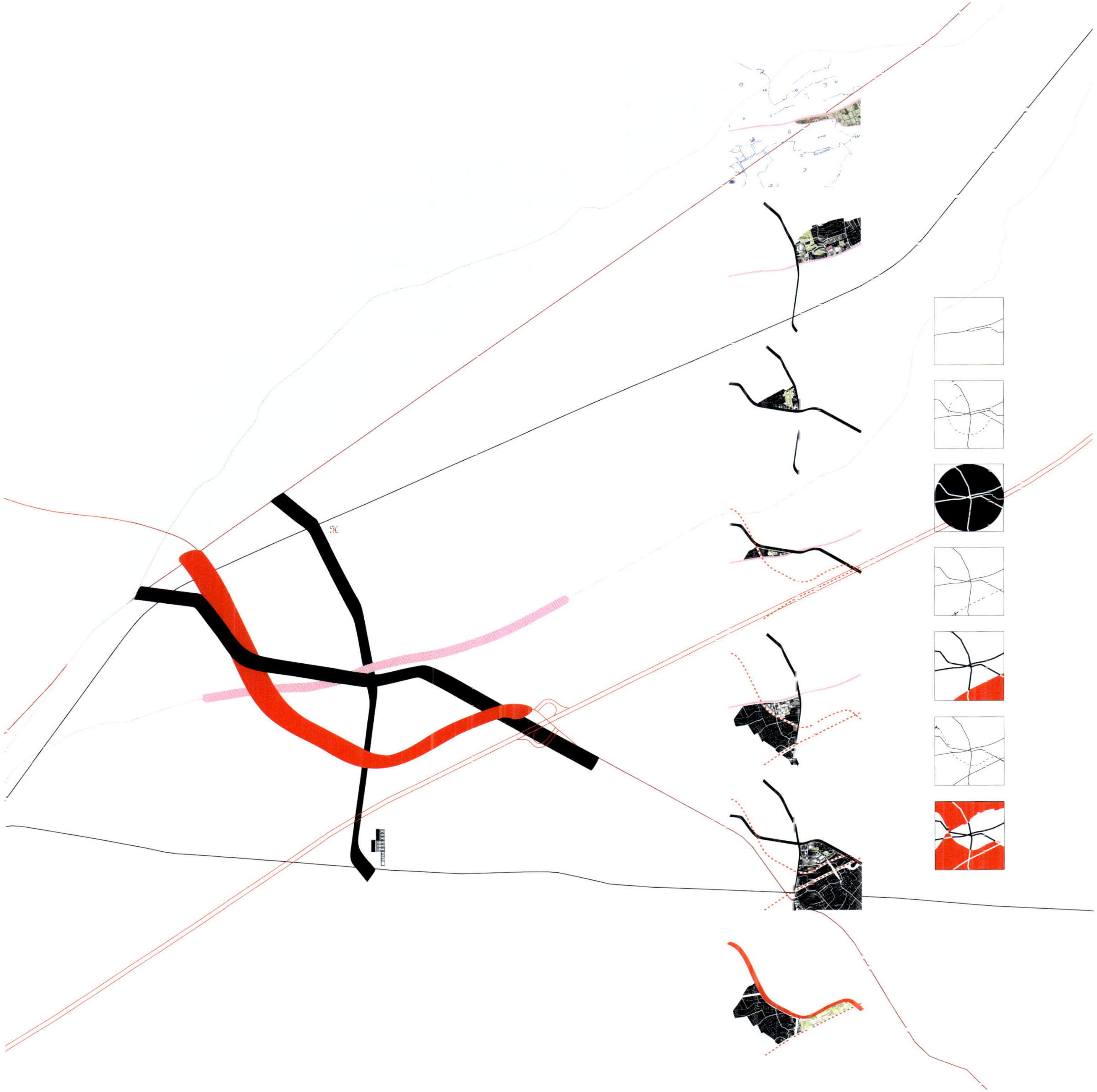

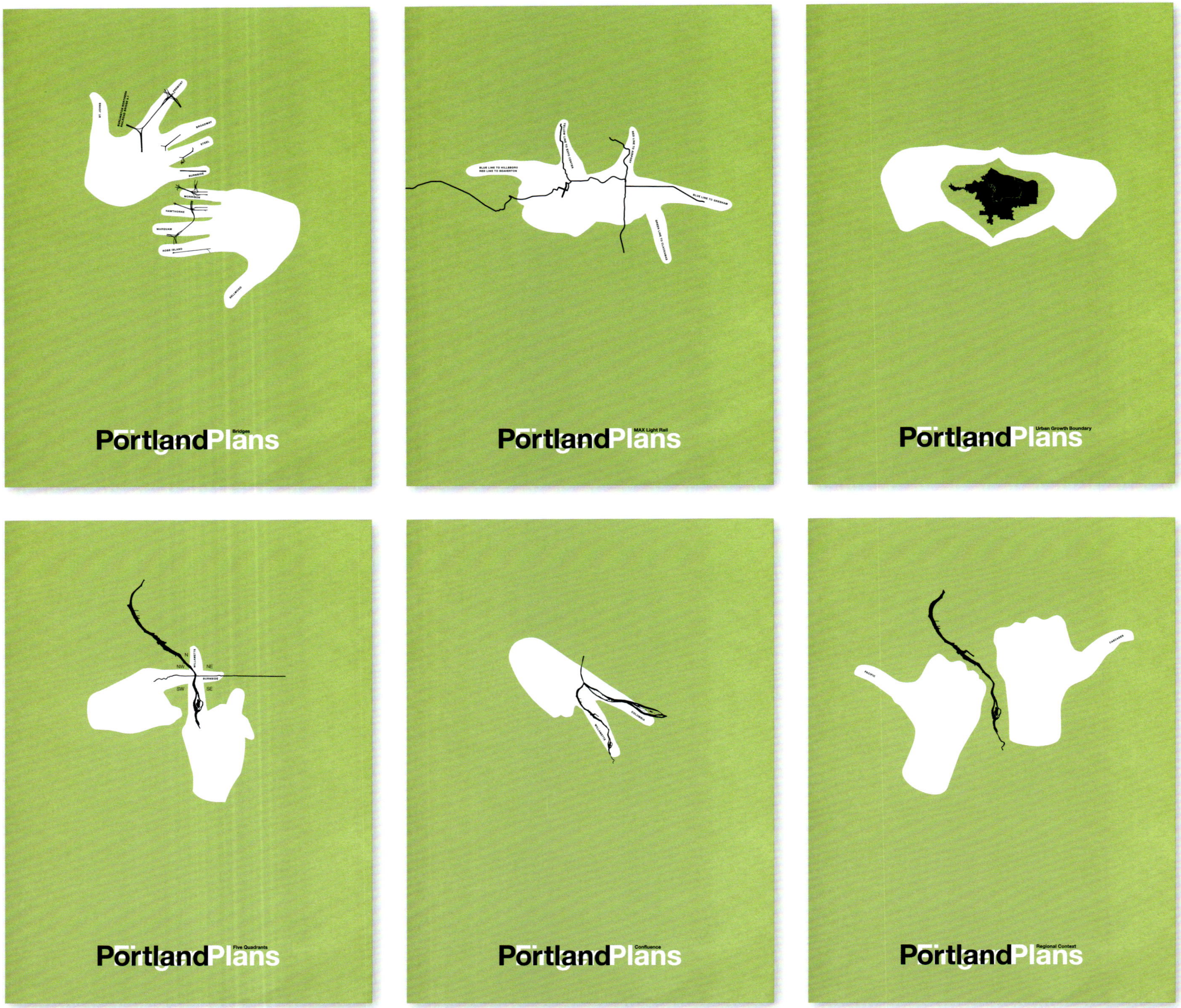
PortlandPlans
Bridges
PortlandPlans
MAX Light Rail
PortlandPlans
Urban Growth Boundary
PortlandPlans
Five Quadrants
N
NW
NE
SW
SE
PortlandPlans
Confluence
PortlandPlans
Regional Context

This project aims to reduce characteristics of Portland's urban form to symbols created with human hands—a relationship that connects a complex system to the human-scaled familiarity of our own bodies.

« Portland Finger Plans

Ryan Sullivan — Paste in Place

This series of six maps of Portland, Oregon, juxtaposes two dramatically different scales: human hands, and the city's urban form. They are inspired by a tradition of using the human body as a metaphor for the city that includes the writings of Vitruvius and da Vinci, and, more recently, the 1947 Copenhagen finger plan.

Cities are large, complex systems composed of different, interrelated components. As such, they are challenging to describe and represent in a single image. This project aims to reduce characteristics of Portland's urban form to symbols created with human hands—a relationship that connects a complex system to the human-scaled familiarity of our own bodies.

In each map, the representation of the human hands and the representation of urban form are rigidly expressed as two different systems in order to clearly articulate their relationship. This idea was reinforced in the decision to use a different color for each system—even in the map titles—and through the printing process, where each layer is screen-printed in a separate pass. In order to emphasize the human body's role in the metaphor, the hands on each map are shown at full-scale so that the reader can experience a 1:1 relationship with the maps.

Each of the six maps describes one of the following geographic aspects of Portland: 1) Bridges, 2) MAX light rail, 3) the Urban Growth Boundary, 4) Portland's five "quadrants" (N, NW, SW, NE, SE), 5) Portland's location at the confluence of the Willamette and Columbia rivers, and 6) Portland's regional location between the Pacific and the Cascades.

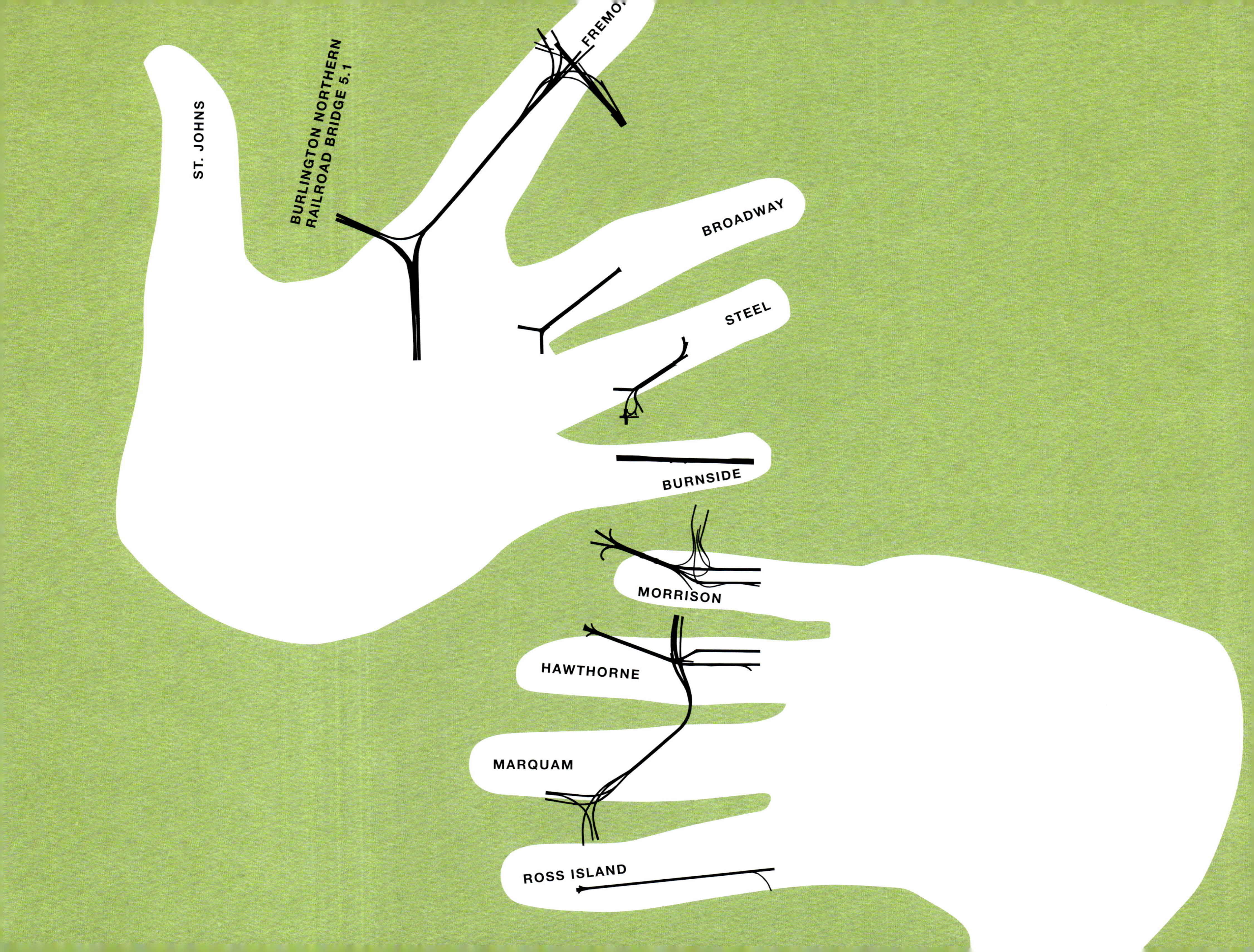
ST. JOHNS
BURLINGTON NORTHERN
RAILROAD BRIDGE 5.1
BROADWAY
STEEL
BURNSIDE
MORRISON
HAWTHORNE
MARQUAM
ROSS ISLAND

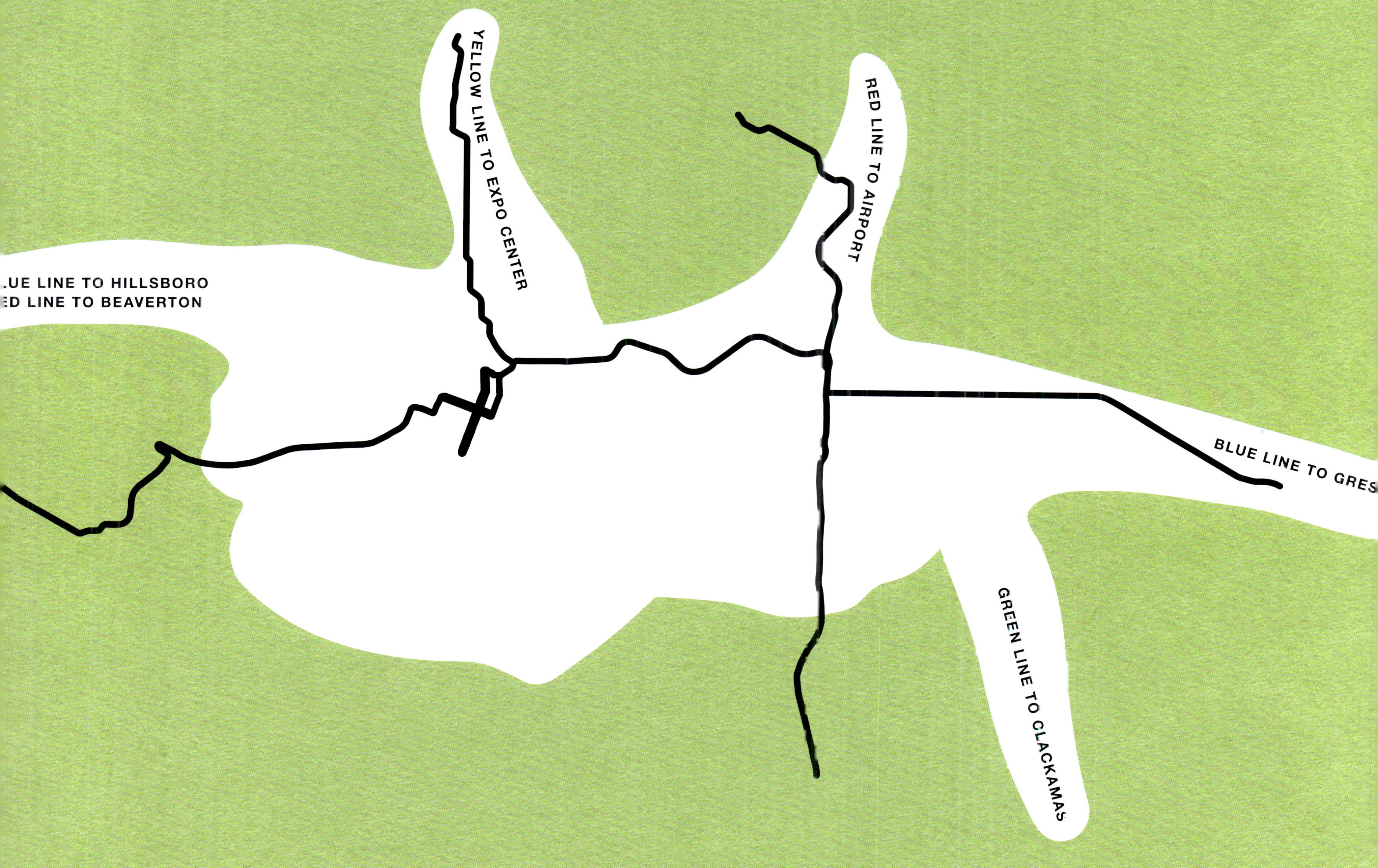
LUE LINE TO HILLSBORO
ED LINE TO BEAVERTON
YELLOW LINE TO EXPO CENTER
RED LINE TO AIRPORT
BLUE LINE TO GRES
GREEN LINE TO CLACKAMAS

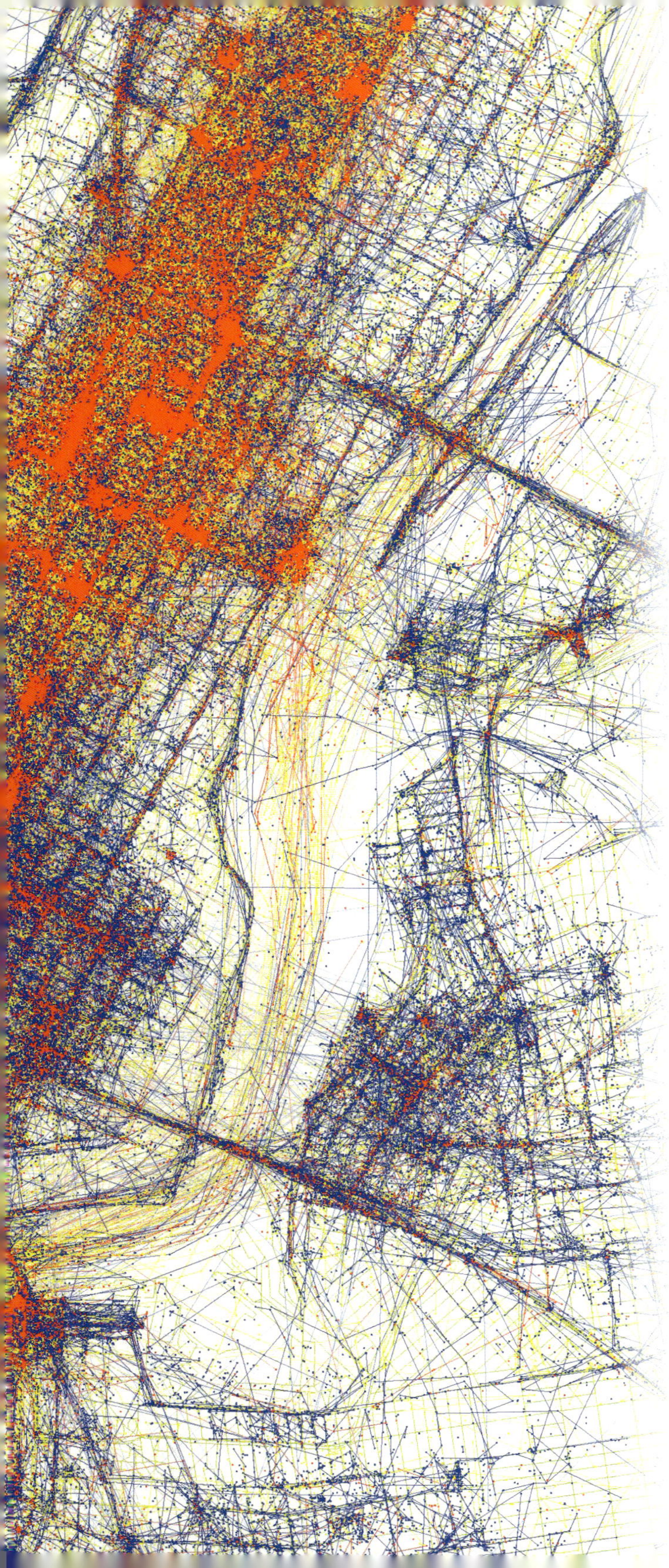

Locals and Tourists »

Eric Fischer

The *Locals and Tourists* maps are the second part of a series that visualizes aspects of a city's structure as they are revealed by the locations and sequence of geotagged photos taken within it and posted to Flickr and Picasa.

I wanted distances and areas within different cities to be easily comparable, so each map is drawn at the same scale: a square 15 miles on each side. This was the largest scale that would allow the largest contiguous area of interest anywhere (e.g., from Prospect Park in Brooklyn to Inwood Park in upper Manhattan) to fit in one map.

The base map is streets and coastlines from OpenStreetMap, drawn as simple lines without labels. Early versions had no base map at all, because for intensely photographed cities like New York, the photo locations alone are sufficient to define the shape of the city. For orientation in smaller cities, however, the base map was necessary. The color of the base map was my second choice after pure yellow, which proved too hard to see.

The photo locations are drawn both as dots representing individual photos, and lines connecting the order in which the photographer visited the locations. I did this in order to give a sense of which locations are significant in and of themselves, and which are important as travel corridors, rather than destinations.

The colors are red (for photos taken by tourists), blue (by locals), and yellow (unsure). Yellow suggested uncertainty, and red suggests the negative feelings that many people associate with tourist-heavy areas. I used blue and avoided green because red and blue provide a much better contrast to my colorblind eyes than red and green do. The red tourist layer is topmost because the message the map is trying to convey is the extent of the places that are important and well known to local people, but unknown to and unvisited by tourists (or, in the case of a few cities like Venice and Las Vegas, the lack of such places).

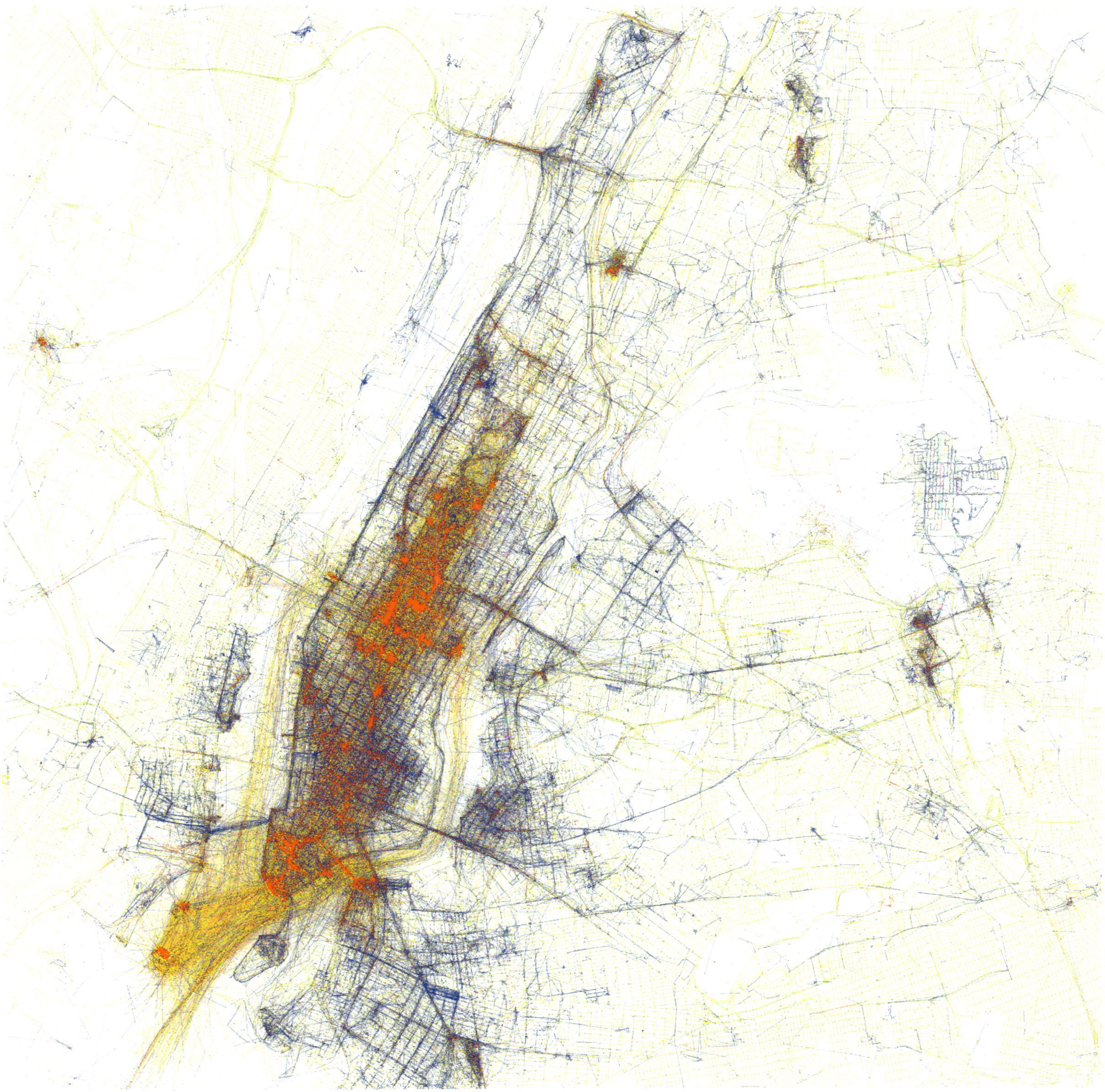

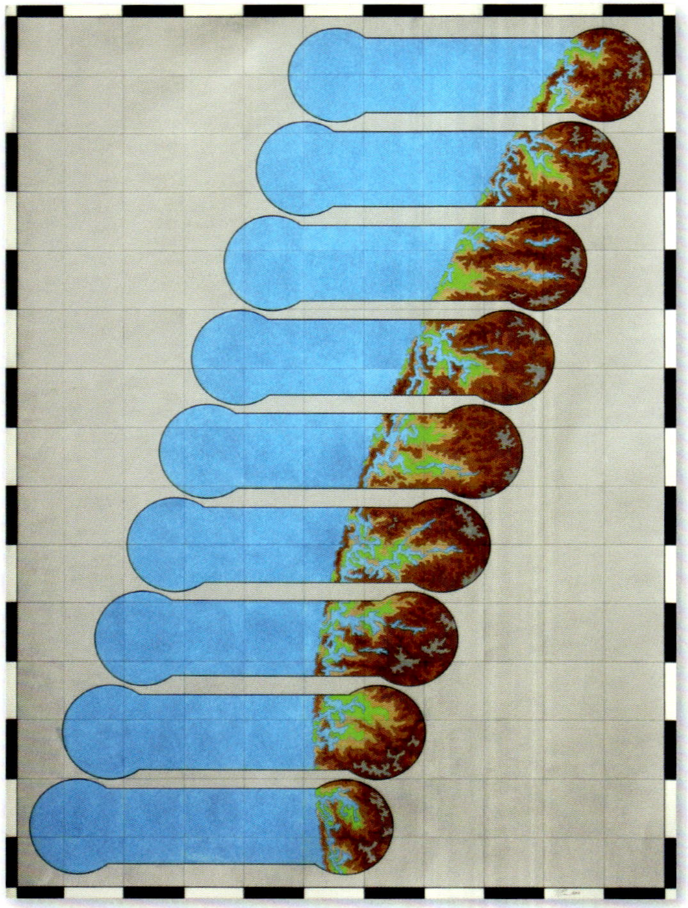

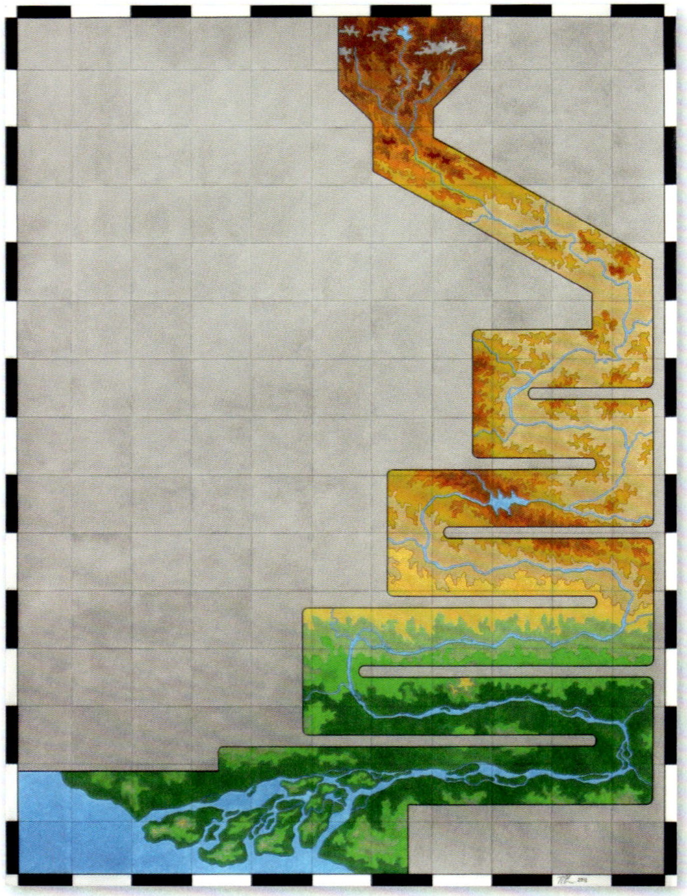

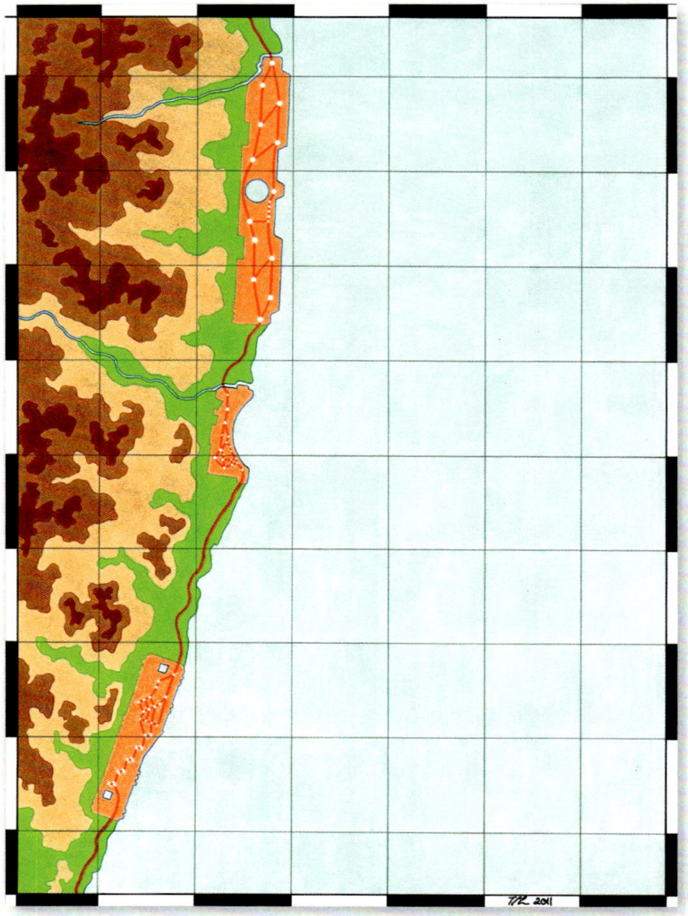

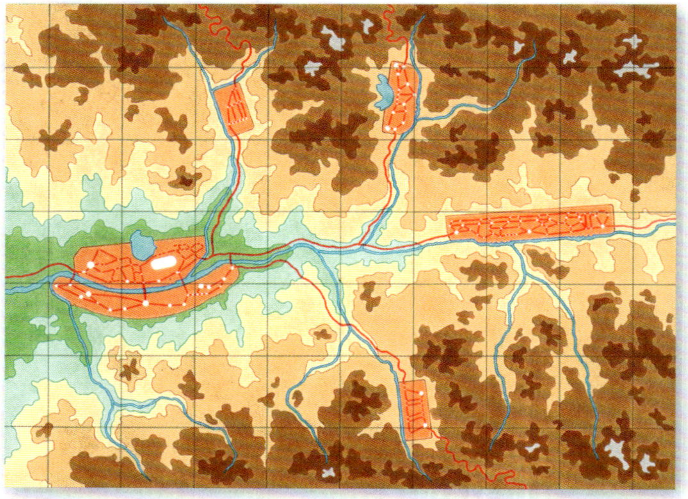

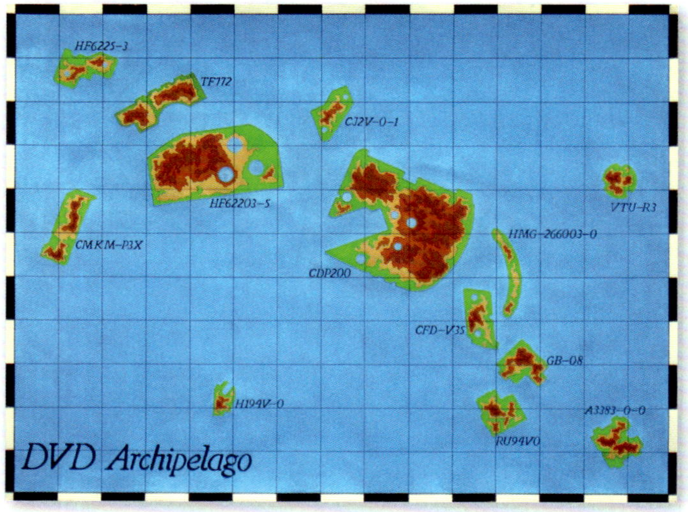
HF6225-3
TF772
CJ2V-0-1
HF62203-S
VTU-R3
CMKM-P3X
HMG-266003-0
CDP200
CFD-V35
GB-08
H194V-0
A3383-0-0
RU94V0
DVD Archipelago

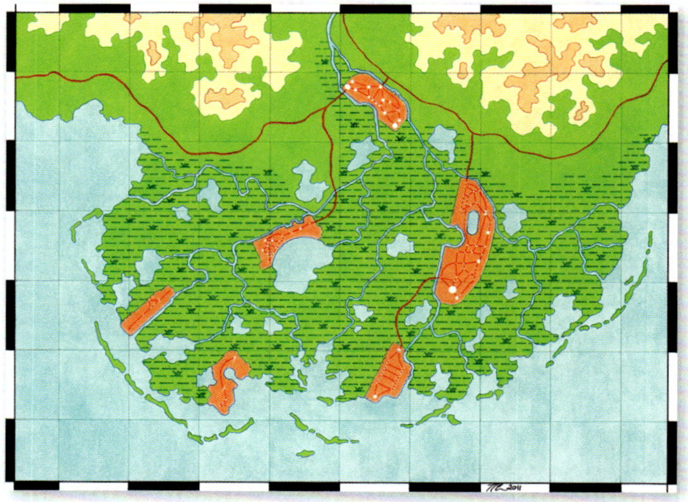

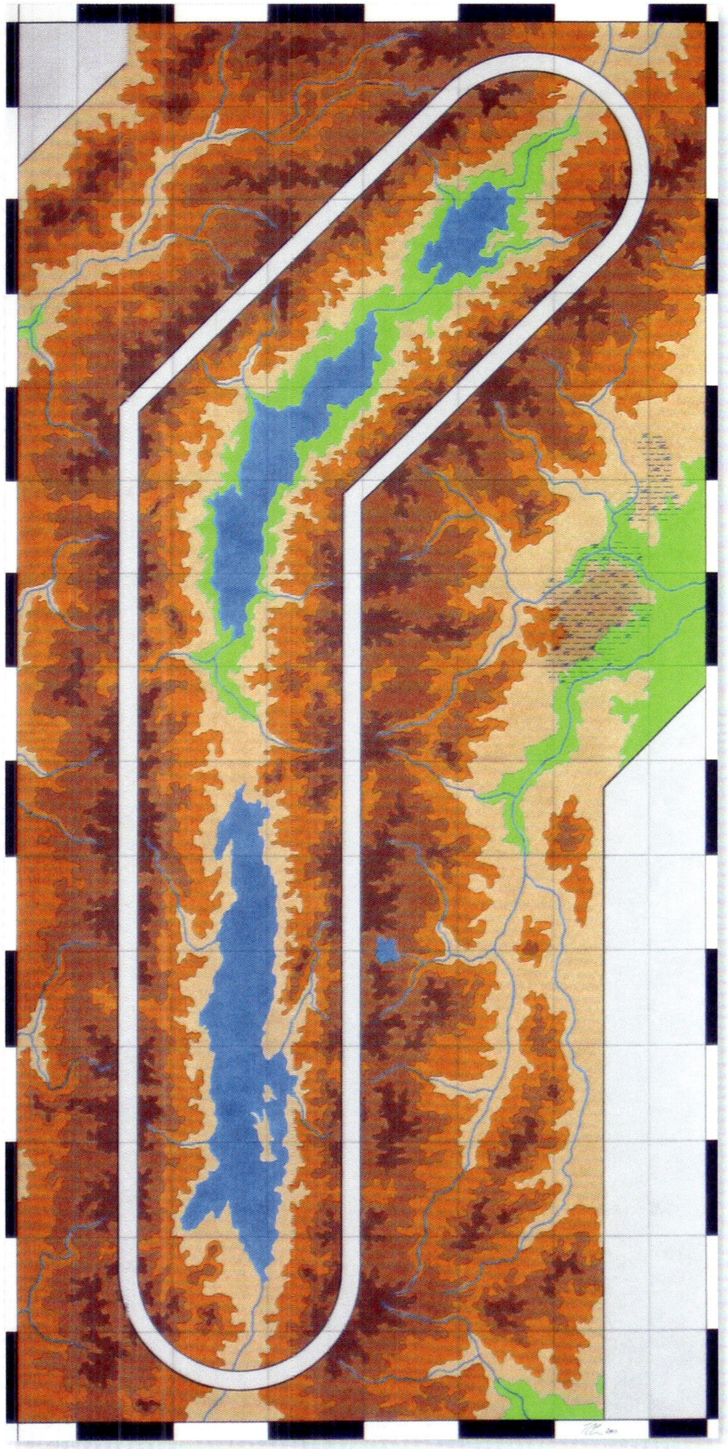

« eCartacacoethes

Tiberiu Chelcea

Opposite page, top row: Rugged Coast, Meandering River, Coastal Cities
Opposite page, bottom row: Cities in a Deep Valley, DVD Archipelago, Delta Cities
This page, left: Rift Valley

The goal of the works in this series is to combine the visual languages of electronic design and maps. To do so, I took advantage of a characteristic found in both printed circuit boards (PCBs) and maps: they are both layered representations of places or objects. PCBs, such as computer motherboards, are made up of wires sandwiched between flat layers of nonconductive materials. Similarly, a map is a layer of information added to a geometrically accurate representation of a place. For example, the shape of an island is represented to scale, and, on top of that, the cartographer adds color-coded topological levels, cities, and roads.

The works in this series use the PCB shapes and the shape of PCB wires as the base layer; adding a topographic layer on top transforms them into maps. For example, in *Rift Valley*, *Meandering River*, and *Rugged Coast*, the wire shapes found on PCBs are a source of inspiration for both the map outlines and the topological features represented in the map. In others, such as *Coastal Cities*, *Delta Cities*, and *Cities in a Deep Valley*, entire circuit boards are letterpress printed, and these shapes become city plans that, in turn, suggest the topology of the rest of the map. In other works, the outlines of several PCBs become islands arranged into archipelagos, as in *DVD Archipelago* (with PCBs rescued from actual DVD players; the factory IDs of the PCBs provide the "island" names). Except for the occasional letterpress-printed PCB, all these maps are hand drawn using ink markers and ink. The maps vary in size from 11" × 15" to 22" × 30". All works were completed in 2011.

These works blur the boundaries between the man-made and the natural, reimagining the tiniest of engineered components as vast land masses; and they illuminate the connections between the ancient art of mapmaking and the very contemporary art of electronic design.

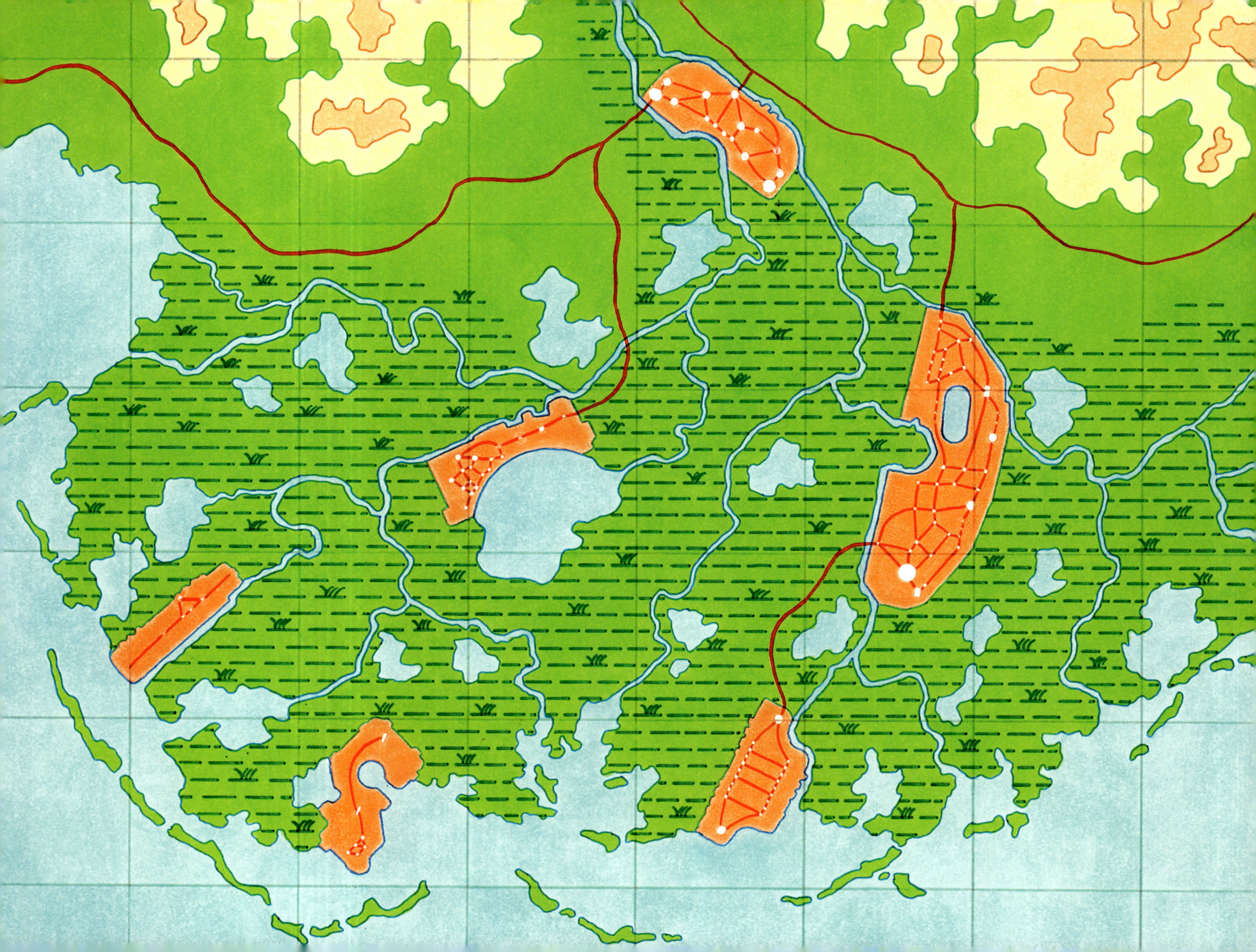

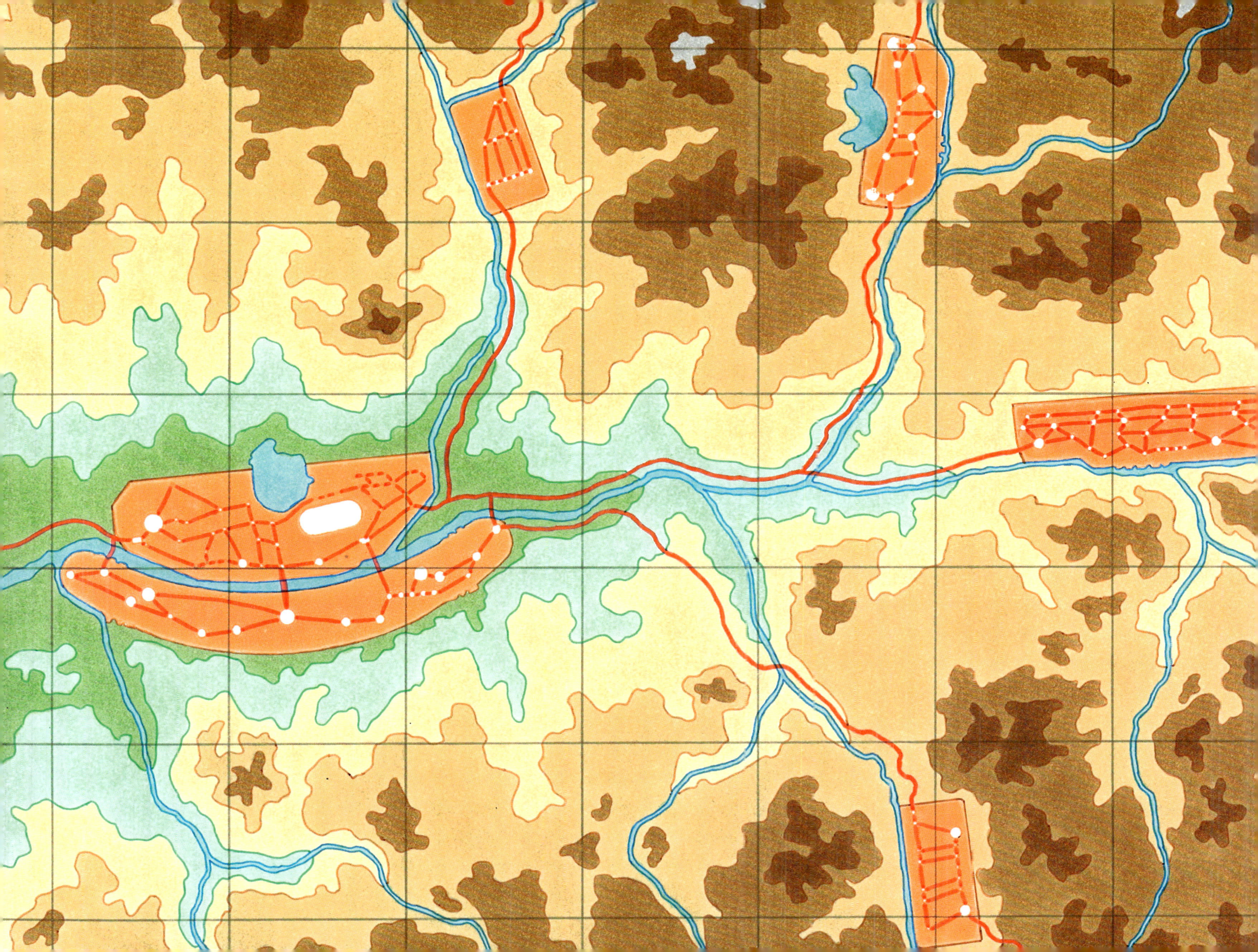

Here Now!: Social Media and the Psychological City »

Sarah Williams, Juan Francisco Saldarriaga

> *Both sites tell us how social media users explore the city, or, more importantly, how they broadcast their exploits.*

Social media have given us a different vantage point from which to understand, explore, navigate, and geographically record the places we live. Sites such as Foursquare and Facebook allow us to spatially mark our explorations in the city, creating rich databases that hold digital imprints of our interactions there.

To analyze these traces, the Foursquare and Facebook Application Programming Interfaces (APIs) were used to access location-based data to determine where social media users broadcast that they are "Here Now." Analysis of this geographic data exposed the psycho-geography and economic terrain of New York City's social media users.

The motivations behind these broadcasts vary between the two social media sites. Foursquare users tell us more about the mundane nuances of life—where their bed is, where they get their morning coffee, or where they work. Facebook users tend to use the site to brag about iconic places they have been to: Times Square, Little Italy, or the Empire State Building. Both sites tell us how social media users explore the city, or, more importantly, how they broadcast their exploits.

Representing the Foursquare and Facebook check-in data as topographic density maps provided a strategy for understanding what the overwhelming number of point-based, check-in data meant in the aggregate. Points that were once mapped on top of each other, obscuring their message, were transformed into contour lines that show the number of check-ins per acre. Representing the contours in 3D allowed their height to also tell the story. This visual strategy highlights that Foursquare check-ins are more dispersed—they show all the mundane things we check in to. On the other hand, Facebook check-ins spiked, illustrating those users' interest in broadcasting popular tourist attractions such as Times Square and Yankee Stadium. The black canvas upon which the contour lines are drawn creates a digital feeling, while also creating a contrast that makes the data the most important element on the map.

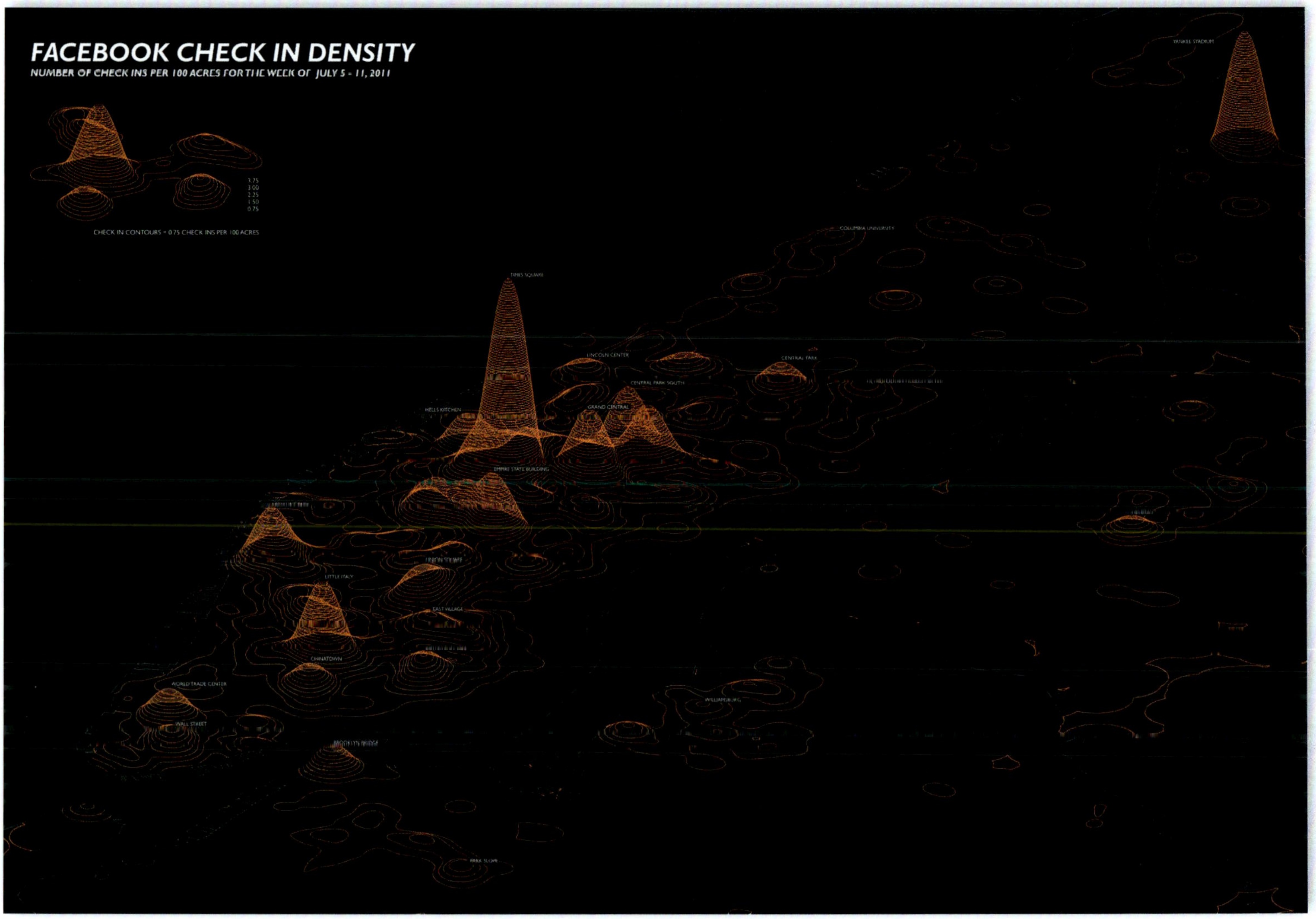
FACEBOOK CHECK IN DENSITY
NUMBER OF CHECK INS PER 100 ACRES FOR THE WEEK OF JULY 5 - 11, 2011
3.75
3.00
2.25
1.50
0.75
CHECK IN CONTOURS = 0.75 CHECK INS PER 100 ACRES
YANKEE STADIUM
COLUMBIA UNIVERSITY
TIMES SQUARE
LINCOLN CENTER
CENTRAL PARK
CENTRAL PARK SOUTH
HELLS KITCHEN
GRAND CENTRAL
EMPIRE STATE BUILDING
LITTLE ITALY
EAST VILLAGE
CHINATOWN
WORLD TRADE CENTER
WALL STREET
BROOKLYN BRIDGE
WILLIAMSBURG
PARK SLOPE

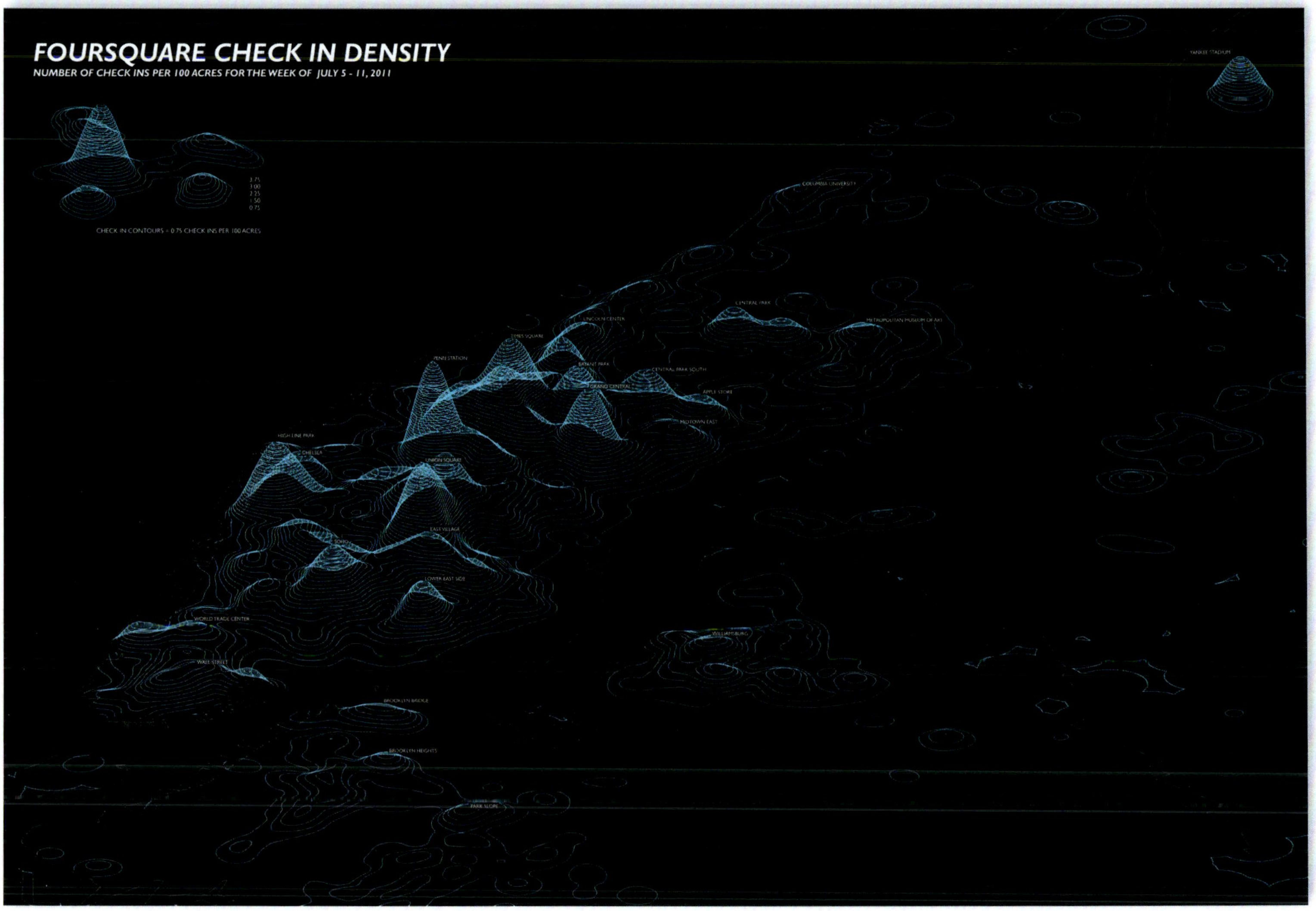
FOURSQUARE CHECK IN DENSITY
NUMBER OF CHECK INS PER 100 ACRES FOR THE WEEK OF JULY 5 - 11, 2011
3.75
3.00
2.25
1.50
0.75
CHECK IN CONTOURS = 0.75 CHECK INS PER 100 ACRES
YANKEE STADIUM
COLUMBIA UNIVERSITY
CENTRAL PARK
METROPOLITAN MUSEUM OF ART
LINCOLN CENTER
TIMES SQUARE
PENN STATION
BRYANT PARK
CENTRAL PARK SOUTH
GRAND CENTRAL
APPLE STORE
MIDTOWN EAST
HIGH LINE PARK
CHELSEA
UNION SQUARE
EAST VILLAGE
SOHO
LOWER EAST SIDE
WORLD TRADE CENTER
WALL STREET
WILLIAMSBURG
BROOKLYN BRIDGE
BROOKLYN HEIGHTS
PARK SLOPE

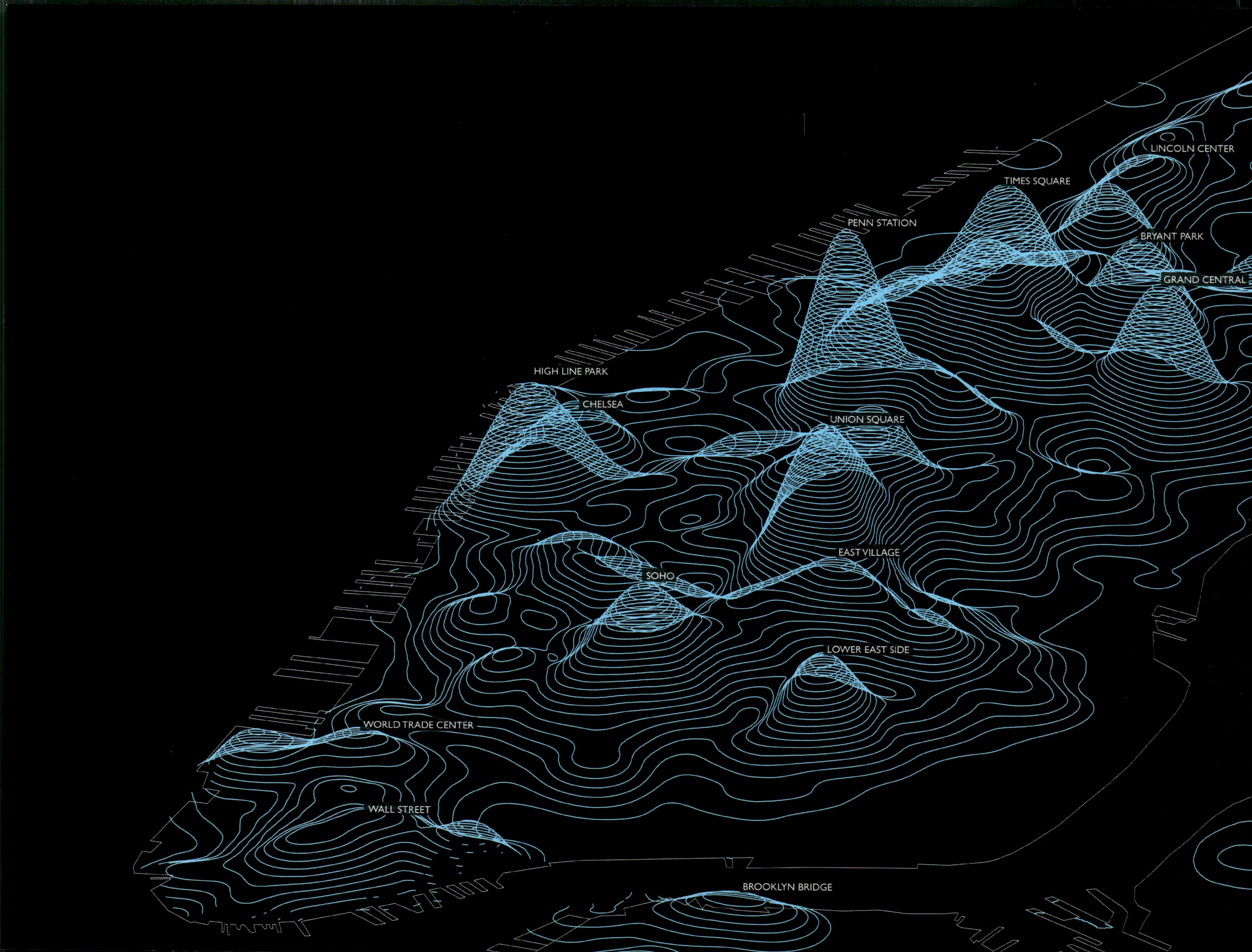
LINCOLN CENTER
TIMES SQUARE
PENN STATION
BRYANT PARK
GRAND CENTRAL
HIGH LINE PARK
CHELSEA
UNION SQUARE
EAST VILLAGE
SOHO
LOWER EAST SIDE
WORLD TRADE CENTER
WALL STREET
BROOKLYN BRIDGE

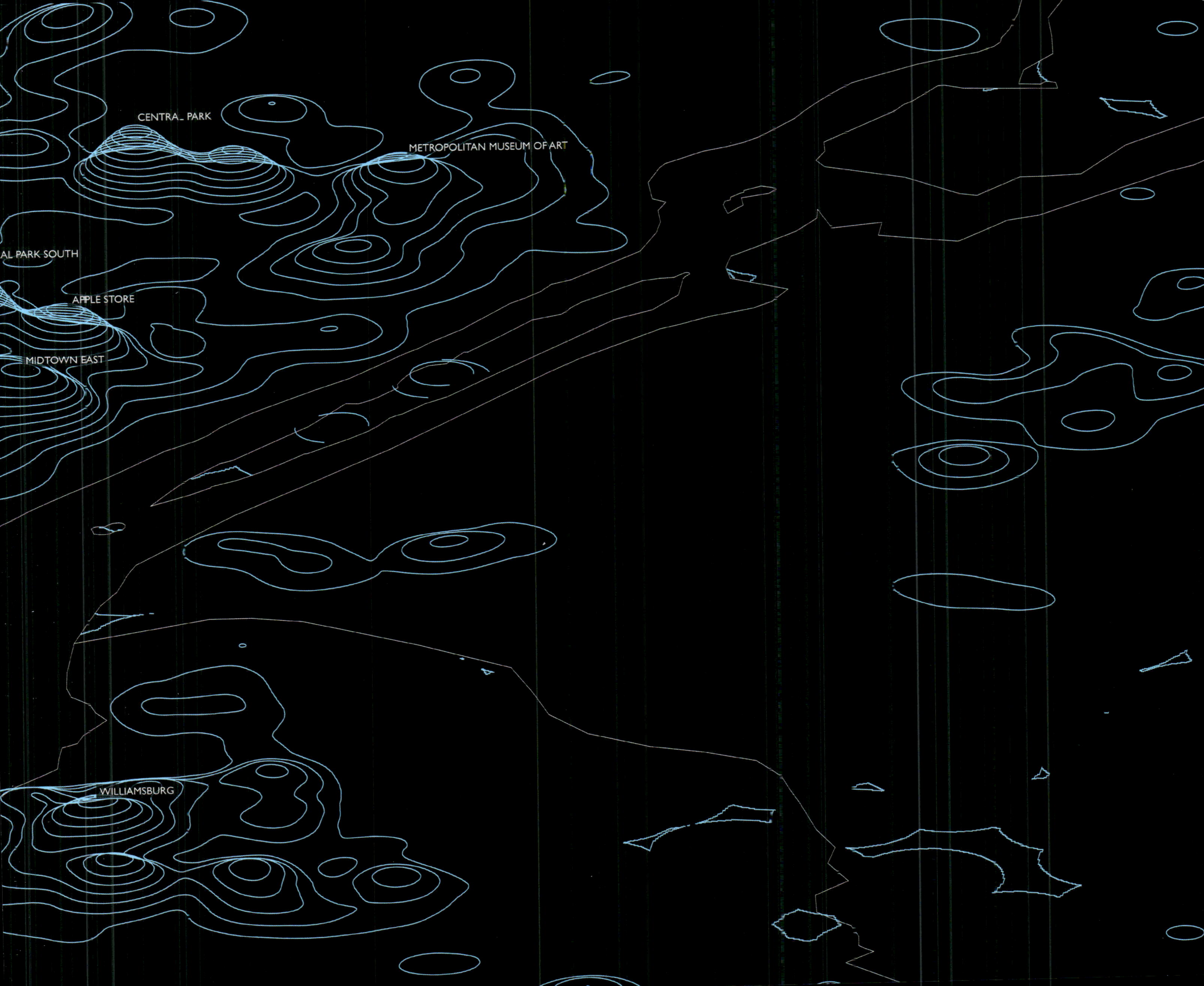
CENTRA_ PARK
METROPOLITAN MUSEUM OF ART
AL PARK SOUTH
APPLE STORE
MIDTOWN EAST
WILLIAMSBURG

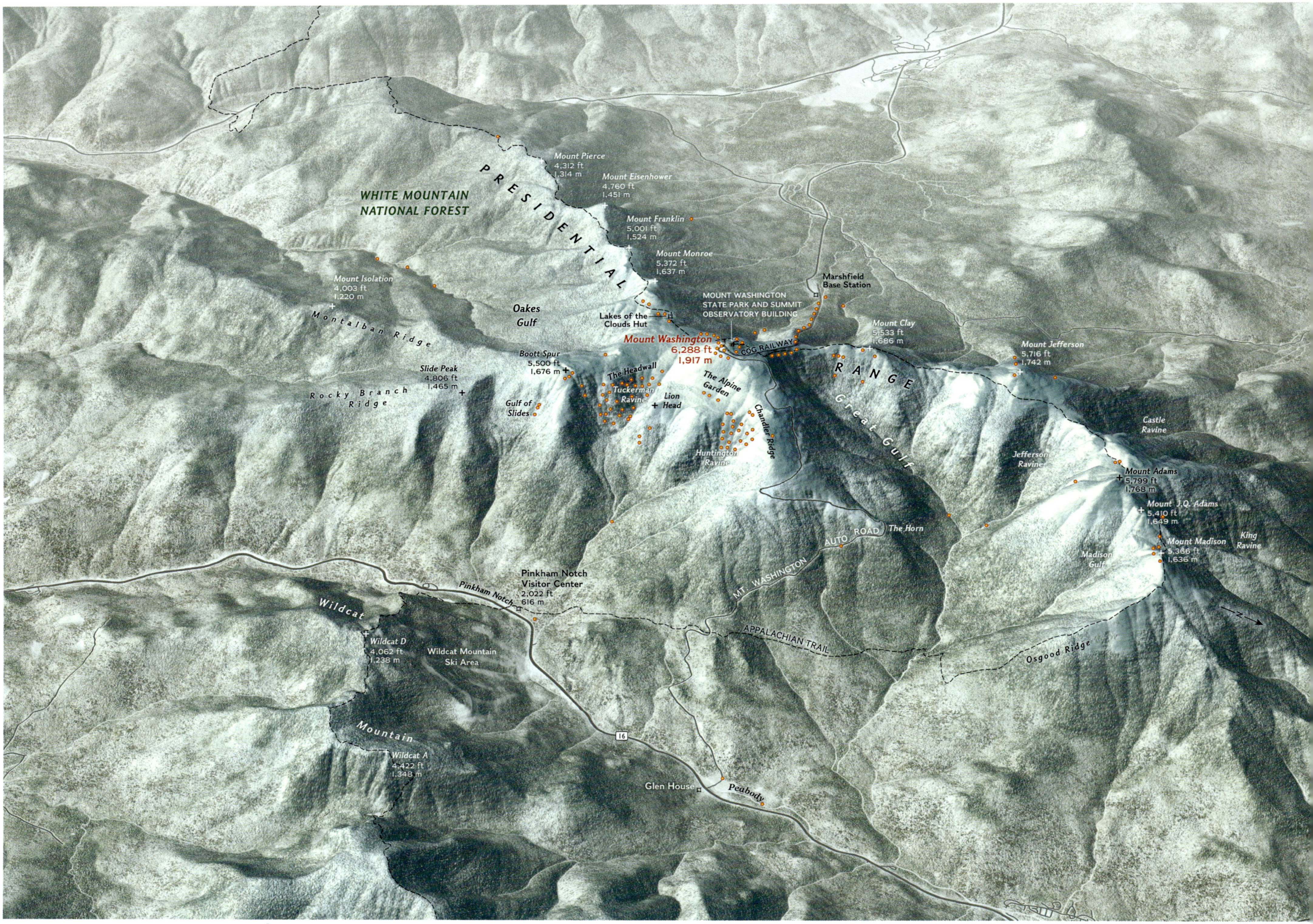
WHITE MOUNTAIN NATIONAL FOREST
PRESIDENTIAL
RANGE
Mount Pierce 4,312 ft 1,314 m
Mount Eisenhower 4,760 ft 1,451 m
Mount Franklin 5,001 ft 1,524 m
Mount Monroe 5,372 ft 1,637 m
Mount Isolation 4,003 ft 1,220 m
Montalban Ridge
Oakes Gulf
Lakes of the Clouds Hut
Marshfield Base Station
MOUNT WASHINGTON STATE PARK AND SUMMIT OBSERVATORY BUILDING
Mount Washington 6,288 ft 1,917 m
COG RAILWAY
Mount Clay 5,533 ft 1,686 m
Mount Jefferson 5,716 ft 1,742 m
Boott Spur 5,500 ft 1,676 m
Slide Peak 4,806 ft 1,465 m
Rocky Branch Ridge
The Headwall
Tuckerman Ravine
Lion Head
The Alpine Garden
Gulf of Slides
Chandler Ridge
Huntington Ravine
Great Gulf
Castle Ravine
Jefferson Ravine
Mount Adams 5,799 ft 1,768 m
Mount J.Q. Adams 5,410 ft 1,649 m
Mount Madison 5,366 ft 1,636 m
King Ravine
Madison Gulf
The Horn
MT. WASHINGTON AUTO ROAD
Pinkham Notch Visitor Center 2,022 ft 616 m
Pinkham Notch
APPALACHIAN TRAIL
Osgood Ridge
Wildcat
Wildcat D 4,062 ft 1,238 m
Wildcat Mountain Ski Area
Mountain
Wildcat A 4,422 ft 1,348 m
16
Glen House
Peabody
N

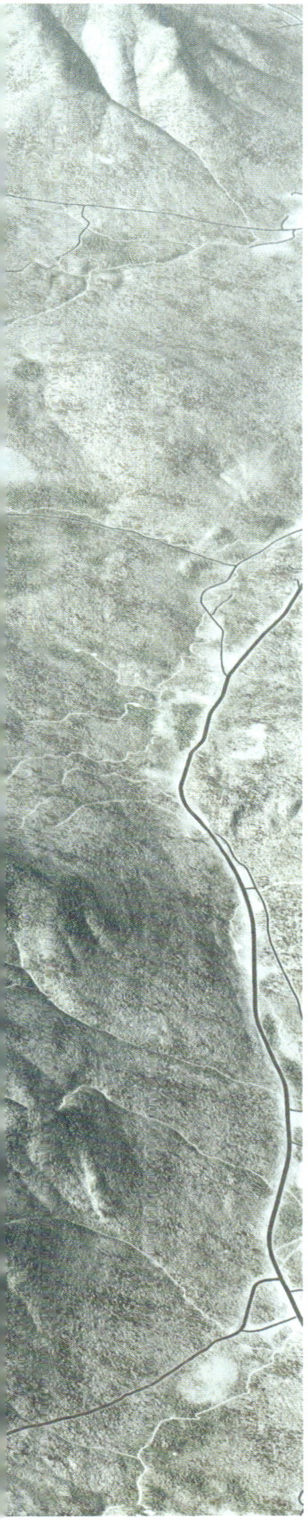

« Panoramic Map of Mount Washington

Alex Tait, Judith Nielsen — International Mapping; *Martin Gamache* — National Geographic Magazine

This panoramic map of Mount Washington helped illustrate an article in *National Geographic Magazine* about the harsh and sometimes dangerous conditions on the highest peak in the northeastern United States. In my design process I carefully manipulated color and composition to help tell the story. The primary subject matter of the map, locations of all deaths on the mountain from 1849 until 2008, compelled me to push the color into a more somber space than most maps I have designed.

I used shades of icy blue and dark spruce and fir greens to set a wintry mood. There is very little variation in hue but quite a lot of variation in value—the dark shadows and snow-white highlights emphasize the harshness of winter in the New England mountains. New Hampshire forests are a combination of deciduous and evergreen coniferous trees. To evoke this patchwork in a winter scene, I used textured areas of rough, spiky tree tips in blue-greens poking through white, interspersed with bare areas of leafless trees with a lightly roughened surface.

To further heighten the contrast of lights and darks, the highest sunlit slopes and the valley streambeds are almost pure white, and the shadowed slopes are set in a very dark green-blue. Use of stark white and black text for the labels also adds a crispness to the map

The three-dimensional panoramic view is a crucial component of this map. It invites the reader to enter the scene in a more visceral way, and perhaps feel some of the icy chill that caused so many of the deaths. I used an oblique view that allows the reader to look into this scene and capture the full drama of winter and danger in one glance. On closer inspection of the view and its pattern of red dots, one can see the individual theaters of death in the alpine cirques on the eastern slopes of Mount Washington and other peaks in the Presidential Range.

» *I wanted the map to look fresh, and I settled on an analogous palette to best impart a sense of relaxation, calm, and peace.*

Marlborough Sounds, New Zealand »

Roger Smith — Geographx

This is a topographic map, designed to highlight and emphasize the intricate landforms, convoluted coastline, and inshore bathymetry of the Marlborough Sounds in New Zealand. This area is a drowned river valley system accessed mostly by boat. With few roads, it is a sparsely settled, popular outdoor recreational playground with holiday homes (baches) dotted around the coast. My primary objective was to create a wall map of the region that is an accurate and functional reference resource for use by residents and visitors alike.

To add interest, the map also provides storyboard interpretation of past events in this historically significant area. During the 1770s, Captain James Cook used Queen Charlotte Sound as a base to explore the South Pacific and Southern Ocean. The map details his many visits, and tells of maritime events associated with the subsequent wave of European explorers, whalers, sealers, and settlers in the area. There are tales, too, of 19th-century Maori inter-tribal conflict, invasion, and conquest.

The map had to be a visually attractive work of art if people were going to buy it and hang it on their walls. I looked for elegance in the choice of typeface, and color was always going to be critical. I wanted the map to look fresh, and I settled on an analogous palette to best impart a sense of relaxation, calm, and peace. These are qualities synonymous with stress-free holidays and also words frequently used to describe the Marlborough Sounds environment during periods of settled, fine weather. I particularly liked the inherent dichotomy of painting a visually harmonious and peaceful-looking picture that contrasted so starkly with the scenes of violence, mayhem, and lawlessness described in many of the accompanying storyboards. You see only the nice; you have to go look for the nasty. The bloody bits may be interesting, but they are confined to small print.

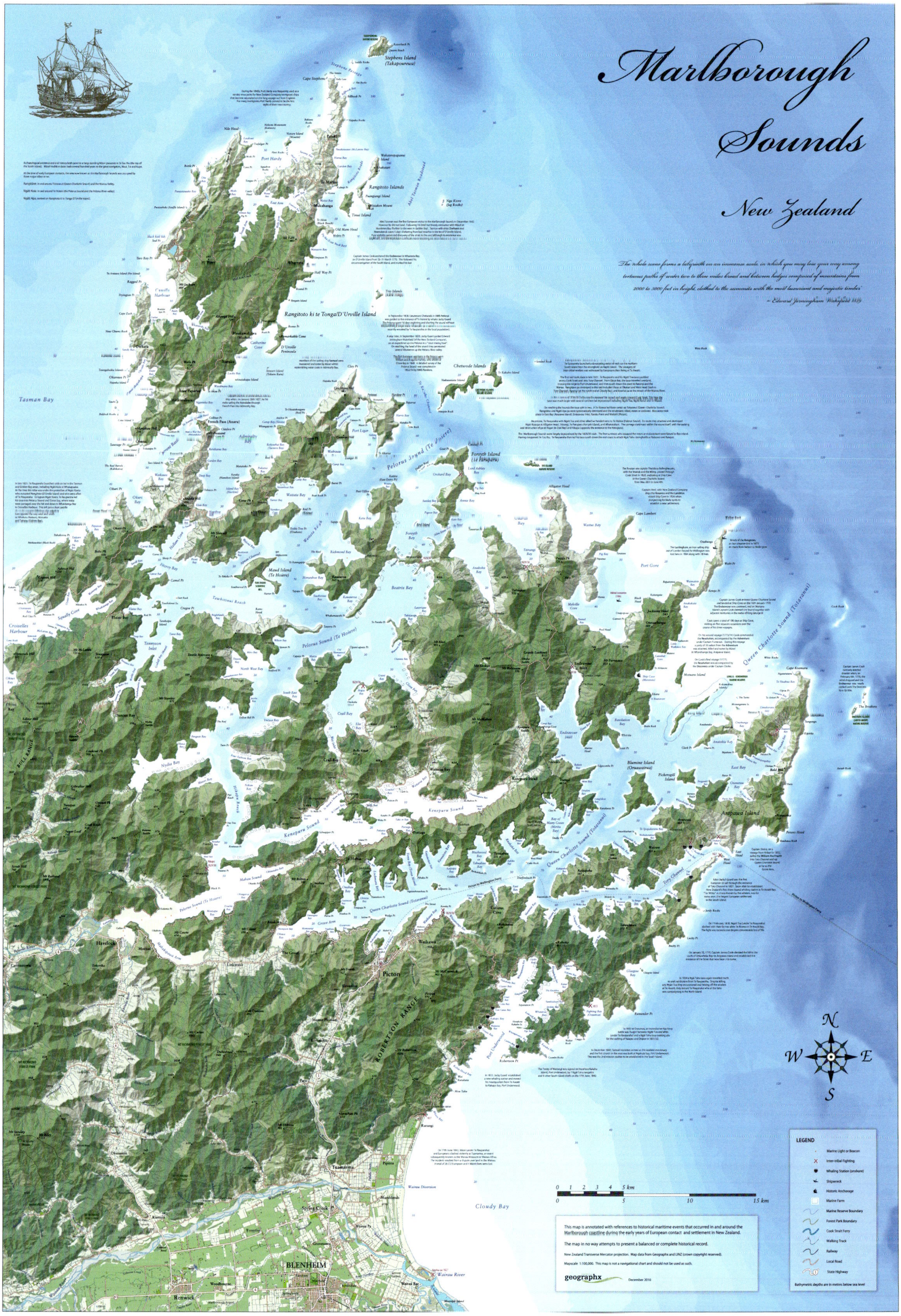

Marlborough Sounds
New Zealand
'The whole scene forms a labyrinth on an immense scale, in which you may lose your way among tortuous paths of water two to three miles broad and between hedges composed of mountains from 2000 to 3000 feet in height, clothed to the summits with the most luxuriant and majestic timber'
– Edward Jerningham Wakefield 1839
Stephens Island (Takapourewa)
Rangitoto ki te Tonga/D'Urville Island
Tasman Bay
Pelorus Sound (Te Hoiere)
Kenepuru Sound
Queen Charlotte Sound (Totaranui)
Arapaoa Island
Blumine Island (Oruawairua)
Forsyth Island (Te Paruparu)
Maud Island (Te Hoiere)
Chetwode Islands
Port Gore
Cloudy Bay
Havelock
Picton
Blenheim
Robertson Range
Wairau River
Tory Channel
0 1 2 3 4 5 km
0 5 10 15 km
This map is annotated with references to historical maritime events that occurred in and around the Marlborough coastline during the early years of European contact and settlement in New Zealand.
The map in no way attempts to present a balanced or complete historical record.
New Zealand Transverse Mercator projection. Map data from Geographx and LINZ (crown copyright reserved).
Mapscale 1:100,000. This map is not a navigational chart and should not be used as such.
geographx
December 2010
LEGEND
Marine Light or Beacon
Inter-tribal Fighting
Whaling Station (onshore)
Shipwreck
Historic Anchorage
Marine Farm
Marine Reserve Boundary
Forest Park Boundary
Cook Strait Ferry
Walking Track
Railway
Local Road
State Highway
Bathymetric depths are in metres below sea level
N
S
E
W

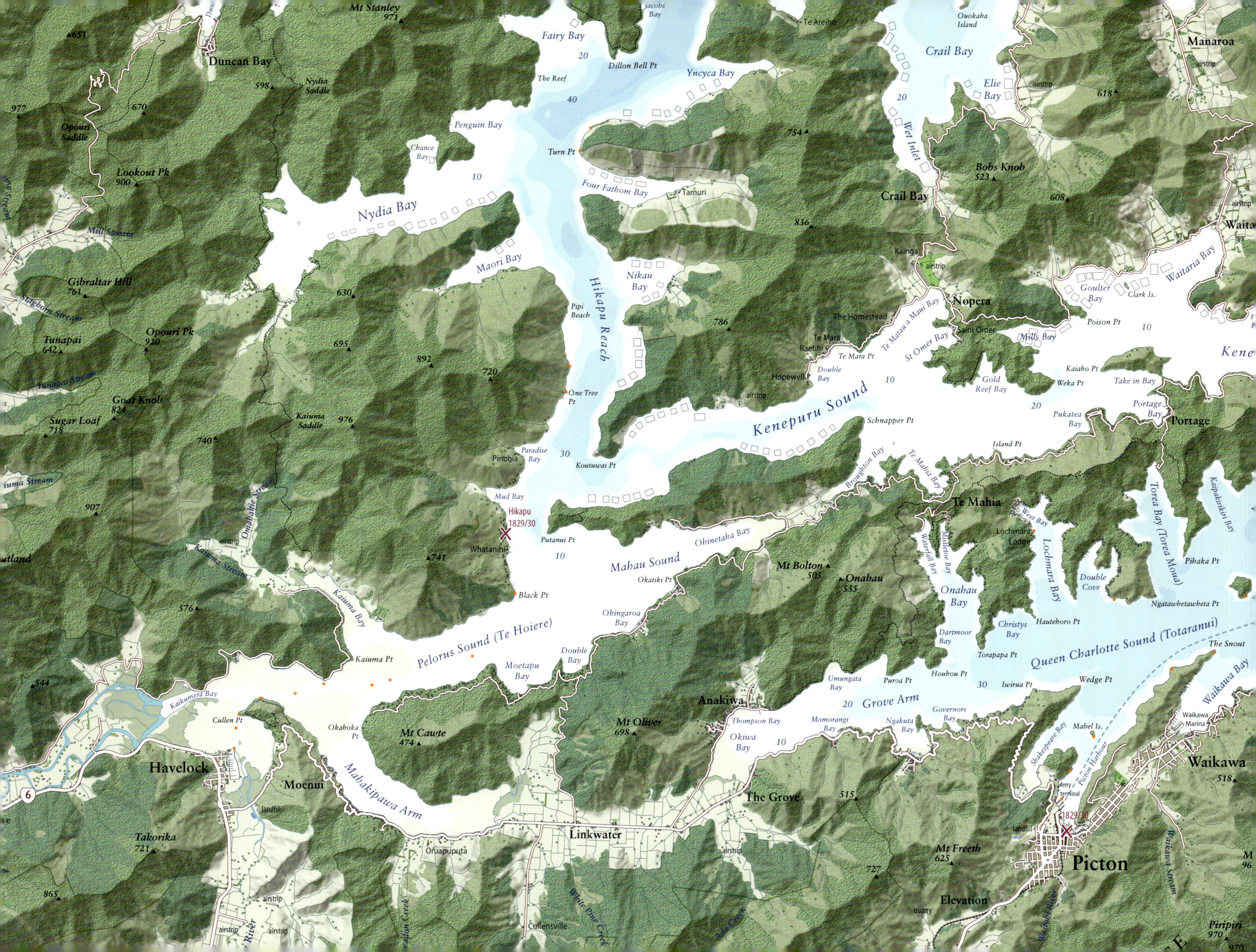

Mt Stanley
971
Jacobs Bay
Te Areihe
Ouokaha Island
Manaroa
Fairy Bay
Crail Bay
Duncan Bay
Dillon Bell Pt
Yncyca Bay
Elie Bay
Nydia Saddle
The Reef
Opouri Saddle
Penguin Bay
Turn Pt
Wet Inlet
Chance Bay
Bobs Knob
523
Lookout Pk
900
Four Fathom Bay
Tamuri
Crail Bay
Nydia Bay
Mill Stream
Maori Bay
Kainga
Waitaria Bay
Nikau Bay
Goulter Bay
Clark Is.
Gibraltar Hill
761
Hikapu Reach
Pipi Beach
Nopera
The Homestead
Te Matau a Maui Bay
Poison Pt
Opouri Pk
920
Saint Omer
Mills Bay
Tunapai
642
Te Mara Raetihi
St Omer Bay
Kene
Te Mara Pt
Double Bay
Kaiabo Pt
Goat Knob
824
One Tree Pt
Hopewell
Gold Reef Bay
Weka Pt
Take in Bay
Sugar Loaf
718
Kaiuma Saddle
Kenepuru Sound
Portage Bay
Portage
Schnapper Pt
Pukatea Bay
Paradise Bay
Pinohia
Koutuwai Pt
Island Pt
Broughton Bay
Te Mahia Bay
Te Mahia
Mud Bay
Hikapu 1829/30
Whatanihi
Putanu Pt
Obinetaba Bay
West Bay
Lochmara Lodge
Torea Bay (Torea Moua)
Kaipikirikiri Bay
Mahau Sound
Waterfall Bay
Mistletoe Bay
Mt Bolton
505
Onahau
555
Lochmara Bay
Pihaka Pt
Okatiki Pt
Black Pt
Onahau Bay
Double Cove
Ngatawhetawheta Pt
Kaiuma Bay
Ohingaroa Bay
Christys Bay
Hauteboro Pt
Pelorus Sound (Te Hoiere)
Dartmoor Bay
Queen Charlotte Sound (Totaranui)
The Snout
Kaiuma Pt
Double Bay
Torapapa Pt
Moetapu Bay
Umungata Bay
Puroa Pt
Houhou Pt
Wedge Pt
Waikawa Bay
Iwirua Pt
Kaikumera Bay
Grove Arm
Anakiwa
Cullen Pt
Governors Bay
Okaboka Pt
Mt Cawte
474
Mt Oliver
698
Thompson Bay
Momorangi Bay
Ngakuta Bay
Mabel Is.
Waikawa Marina
Okiwa Bay
Shakespeare Bay
Picton Harbour
Waikawa
518
Havelock
Moenui
Mahakipawa Arm
ferry terminal
The Grove
515
landfill
Takorika
721
Linkwater
Oruapuputa
1829/30
Mt Freeth
625
Picton
727
Elevation
quarry
Cullensville
White Pine Creek
Waikawa Stream
airstrip
Piripiri
970
6

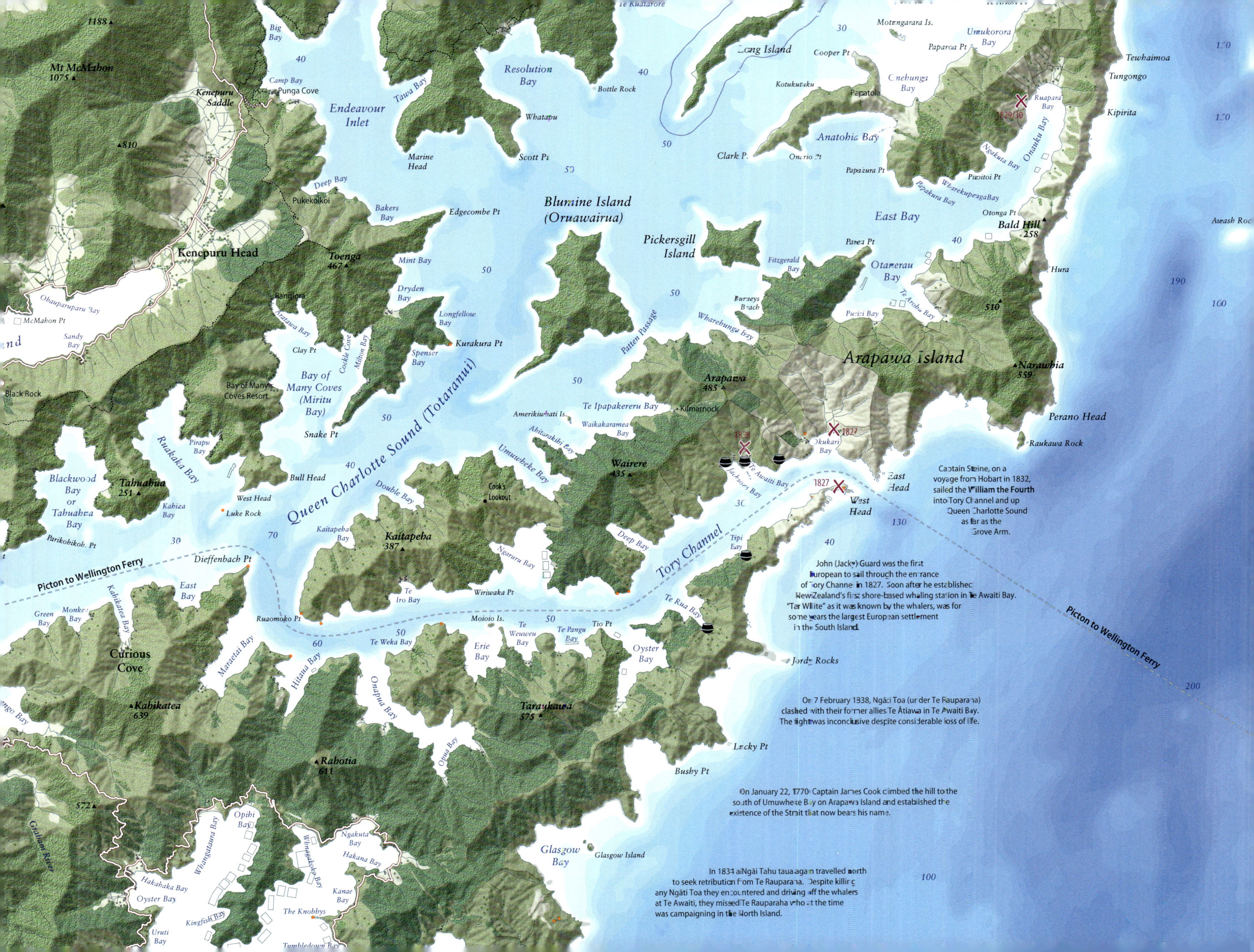

Big Bay
Mt McMahon
1075
1188
Camp Bay
Punga Cove
Kenepuru Saddle
Endeavour Inlet
Resolution Bay
Bottle Rock
Tawa Bay
Whatapu
Scott Pt
Marine Head
Long Island
Cooper Pt
Motungarara Is.
Kotukutuku
Anatohia Bay
Clark Pt
Oneroa Pt
Papatoia Pt
Umukorora Bay
Paparoa Pt
Onehunga Bay
Ruapara Bay
1829/30
Tewhaimoa
Tungongo
Kipirita
Onauku Bay
Ngakuta Bay
Pipitoi Pt
Otonga Pt
Bald Hill
258
Awash Rock
East Bay
Deep Bay
Bakers Bay
Edgecombe Pt
Blumine Island (Oruawairua)
Pickersgill Island
Pukekoikoi
Kenepuru Head
Toenga
467
Mint Bay
Dryden Bay
Longfellow Bay
Kurakura Pt
Rangiora
Ohauparuparu Bay
McMahon Pt
Sandy Bay
Arapawa Island
Arapawa
485
Narawhia
559
510
Hura
Otanerau Bay
Pareu Pt
Fitzgerald Bay
Wharehunga Bay
Patten Passage
Clay Pt
Bay of Many Coves (Miritu Bay)
Bay of Many Coves Resort
Spenser Bay
Cockle Cove
Black Rock
Snake Pt
Queen Charlotte Sound (Totaranui)
Te Ipapakereru Bay
Kilmarnock
Amerikiwhati Is.
Waikakaramea Bay
Okukari Bay
1827
Perano Head
Raukawa Rock
East Head
West Head
Wairere
435
Umuwheke Bay
Cook's Lookout
Double Bay
Blackwood Bay or Tahuahua Bay
Tahuahua
251
Ruakaka Bay
Pirapu Bay
Kahiza Bay
Bull Head
West Head
Luke Rock
Parikobikob Pt
Kaitapeha
387
Kaitapeha Bay
Deep Bay
Tory Channel
Tipi Bay
Te Awaiti Bay
Jacksons Bay
Picton to Wellington Ferry
Dieffenbach Pt
East Bay
Green Bay
Monkey Bay
Kaikatea Bay
Curious Cove
Ruaomoko Pt
Te Iro Bay
Wiriwaka Pt
Moioio Is.
Te Weka Bay
Te Weuweu Bay
Te Pangu Bay
Tio Pt
Erie Bay
Oyster Bay
Te Rua Bay
Jordy Rocks
Maraetai Bay
Hitaua Bay
Onapua Bay
Kahikatea
639
Taraukawa
575
Rahotia
611
Otua Bay
Lucky Pt
Busby Pt
572
Opihi Bay
Whangatoetoe Bay
Ngakuta Bay
Hakana Bay
Hakahaka Bay
Oyster Bay
Kingfish Bay
Uruti Bay
Kanae Bay
The Knobbys
Tumbledown Bay
Glasgow Bay
Glasgow Island
Grahams River
Captain Steine, on a voyage from Hobart in 1832, sailed the William the Fourth into Tory Channel and up Queen Charlotte Sound as far as the Grove Arm.
John (Jacky) Guard was the first European to sail through the entrance of Tory Channel in 1827. Soon after he established New Zealand's first shore-based whaling station in Te Awaiti Bay. "Tar White" as it was known by the whalers, was for some years the largest European settlement in the South Island.
On 7 February 1838, Ngāti Toa (under Te Rauparaha) clashed with their former allies Te Ātiawa in Te Awaiti Bay. The fight was inconclusive despite considerable loss of life.
On January 22, 1770 Captain James Cook climbed the hill to the south of Umuwheke Bay on Arapawa Island and established the existence of the Strait that now bears his name.
In 1834 a Ngāi Tahu taua again travelled north to seek retribution from Te Rauparaha. Despite killing any Ngāti Toa they encountered and driving off the whalers at Te Awaiti, they missed Te Rauparaha who at the time was campaigning in the North Island.

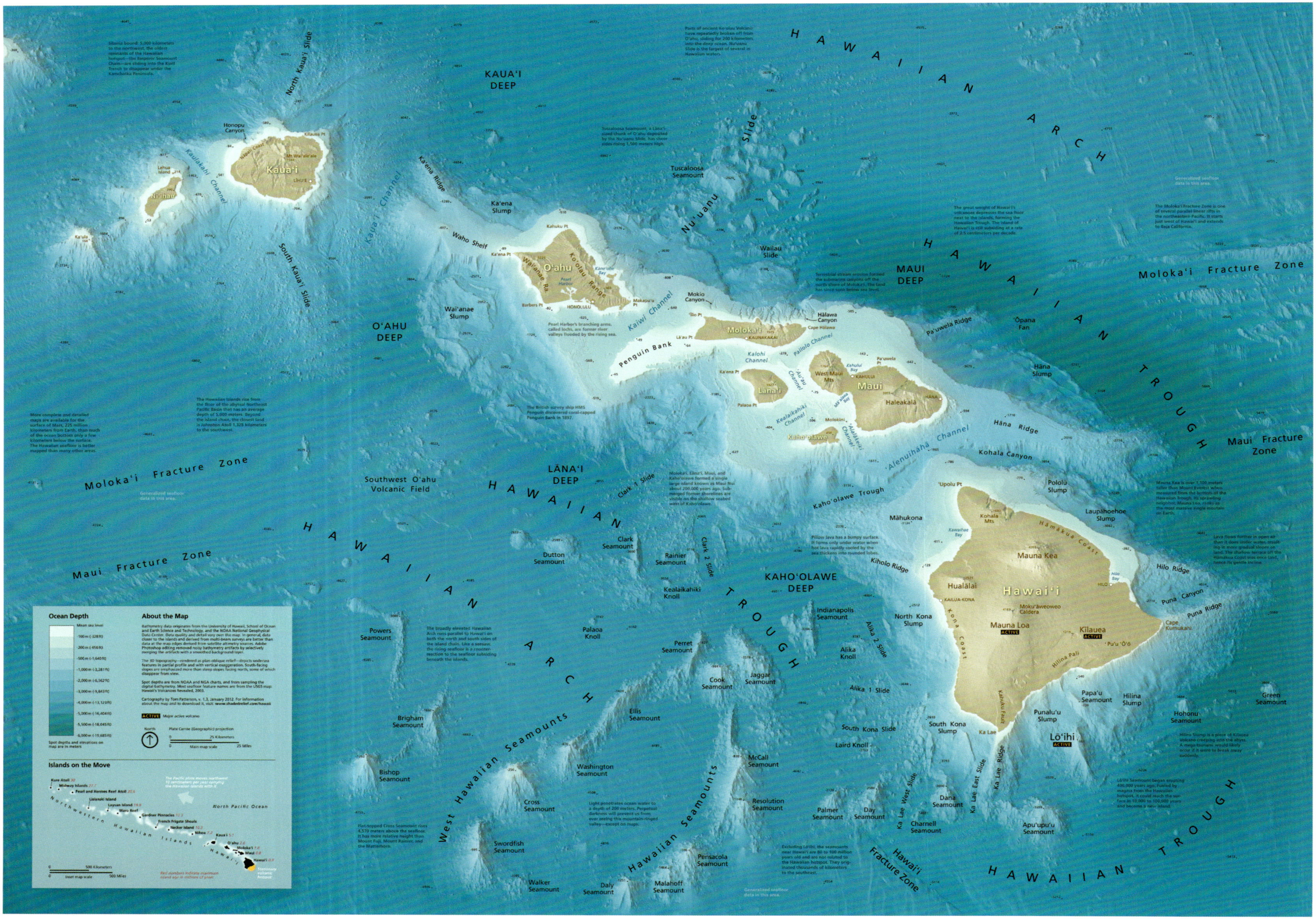

HAWAIIAN ARCH
KAUA'I DEEP
North Kaua'i Slide
Honopu Canyon
Kaua'i
Ni'ihau
Lehua Island
Ka'ula
Kaulakahi Channel
South Kaua'i Slide
Kaua'i Channel
Ka'ena Ridge
Ka'ena Slump
Waho Shelf
O'ahu
Wai'anae Slump
O'AHU DEEP
Tuscaloosa Seamount
Nu'uanu Slide
Wailau Slide
Mokio Canyon
Kaiwi Channel
Penguin Bank
Moloka'i
Halawa Canyon
Pailolo Channel
Kalohi Channel
Lāna'i
Au'au Channel
West Maui Mts
Maui
Haleakalā
Kealaikahiki Channel
Kaho'olawe
'Alalākeiki Channel
MAUI DEEP
HAWAIIAN TROUGH
Pa'uwela Ridge
'Opana Fan
Hāna Slump
Hāna Ridge
Moloka'i Fracture Zone
Maui Fracture Zone
'Alenuihāhā Channel
Kohala Canyon
Pololū Slump
Laupāhoehoe Slump
Hilo Ridge
Puna Canyon
Puna Ridge
Cape Kumukahi
LĀNA'I DEEP
Southwest O'ahu Volcanic Field
Clark 1 Slide
Kaho'olawe Trough
Māhukona
'Upolu Pt
Kohala Mts
Mauna Kea
Hualālai
Hawai'i
Moku'āweoweo Caldera
Mauna Loa
ACTIVE
Kīlauea
ACTIVE
Pu'u 'Ō'ō
Hilina Pali
Kona Coast
Kiholo Ridge
KAHO'OLAWE DEEP
Clark Seamount
Rainier Seamount
Clark 2 Slide
Dutton Seamount
Kealaikahiki Knoll
Indianapolis Seamount
North Kona Slump
Alika 2 Slide
Alika Knoll
Alika 1 Slide
Powers Seamount
Palaoa Knoll
Perret Seamount
Cook Seamount
Jaggar Seamount
Ellis Seamount
Brigham Seamount
Bishop Seamount
Washington Seamount
McCall Seamount
South Kona Slide
South Kona Slump
Laird Knoll
Papa'u Seamount
Hilina Slump
Punalu'u Slump
Hohonu Seamount
Green Seamount
Lō'ihi
ACTIVE
Ka Lae
Ka Lae East Slide
Ka Lae Ridge
Ka Lae West Slide
Dana Seamount
Charnell Seamount
Day Seamount
Palmer Seamount
Resolution Seamount
Apu'upu'u Seamount
Cross Seamount
Swordfish Seamount
Walker Seamount
Daly Seamount
Malahoff Seamount
Pensacola Seamount
West Hawaiian Seamounts
Hawaiian Seamounts
Hawai'i Fracture Zone
Ocean Depth
About the Map
Islands on the Move
North Pacific Ocean
Northwestern Hawaiian Islands

« Seafloor Map of Hawai'i

Tom Patterson — US National Park Service

The volcanic Hawaiian Islands are among Earth's most prominent mountains when measured from their bases on the ocean floor. For example, the summit of Mauna Kea on the island of Hawai'i rises 10,000 meters—a full 1,100 meters taller than Mount Everest's height above sea level. Yet with 60 percent of these islands' total height hidden beneath the Pacific, most people do not realize the significance and enormity of Hawai'i's mountains. The *Seafloor Map of Hawai'i*—designed as a "soft science" reference map for general audiences—attempts to correct this misperception. It also depicts a place that we will never fully see. Light penetrates ocean water to a depth of 200 meters and perpetual darkness cloaks what lies beyond. How, then, should the seafloor appear on a map when we can only imagine what it looks like?

Gauzy blue tints portray the depths progressing from light shallows to dark deep areas. It is a conventional color sequence that mimics how we see deepening water from boats, bridges, or when swimming. There are also graphical advantages to this choice. Because blue is the most visually recessive hue and dark values appear lower, readers will perceive the bathymetry as occupying the lowest areas on a map. Blue depth tints blend harmoniously with gray-shaded relief. It also helps that blue is overwhelmingly the most popular color.

Seafloor topography appears as plan oblique relief, which contains three-dimensional offset—mountains project upwards toward the top of the map and valleys downward. Besides cartographic drama, plan oblique relief offers seafloor maps another advantage. Because the illumination originates from the front left, compared to back left with conventional shaded relief, shadows fall on slopes facing away from the reader. The lighter relief is easier to see against dark blue bathymetric tints in deep areas, improving overall map legibility.

The take away message to readers: most of Hawai'i lies beneath the waves, and only with a map can one visualize the islands' entirety.

How, then, should the seafloor appear on a map when we can only imagine what it looks like?

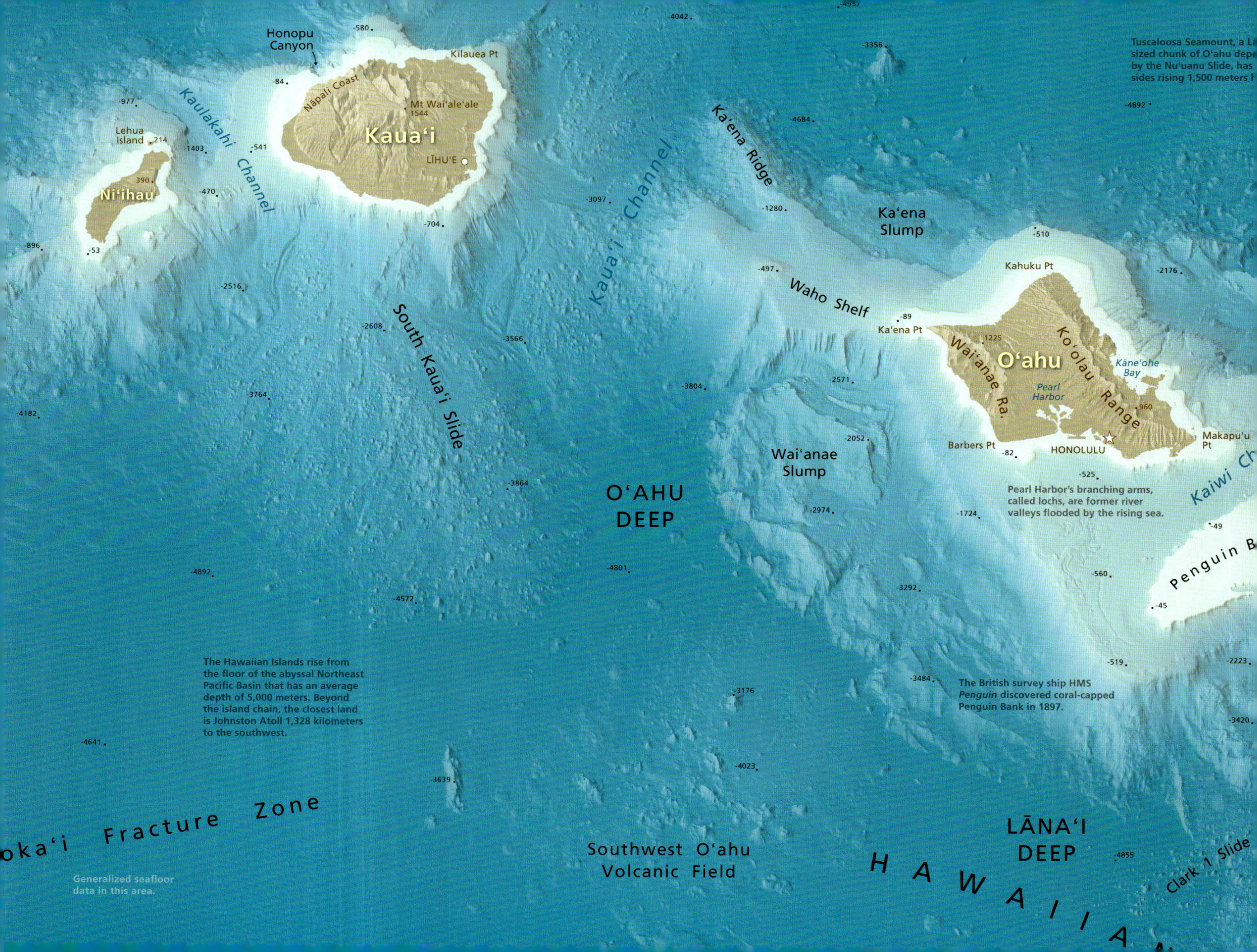

Honopu Canyon
Kilauea Pt
Napali Coast
Mt Wai'ale'ale 1544
Kaua'i
LĪHU'E
Kaulakahi Channel
Lehua Island
Ni'ihau
Kaua'i Channel
Ka'ena Ridge
Ka'ena Slump
Waho Shelf
Ka'ena Pt
Kahuku Pt
Wai'anae Ra.
Ko'olau Range
O'ahu
Kāne'ohe Bay
Pearl Harbor
Barbers Pt
HONOLULU
Makapu'u Pt
Kaiwi Ch
Penguin B
Wai'anae Slump
South Kaua'i Slide
O'AHU DEEP
LĀNA'I DEEP
Clark 1 Slide
H A W A I I A
oka'i Fracture Zone
Southwest O'ahu Volcanic Field
Tuscaloosa Seamount, a L
sized chunk of O'ahu dep
by the Nu'uanu Slide, has
sides rising 1,500 meters
Pearl Harbor's branching arms, called lochs, are former river valleys flooded by the rising sea.
The British survey ship HMS Penguin discovered coral-capped Penguin Bank in 1897.
The Hawaiian Islands rise from the floor of the abyssal Northeast Pacific Basin that has an average depth of 5,000 meters. Beyond the island chain, the closest land is Johnston Atoll 1,328 kilometers to the southwest.
Generalized seafloor data in this area.

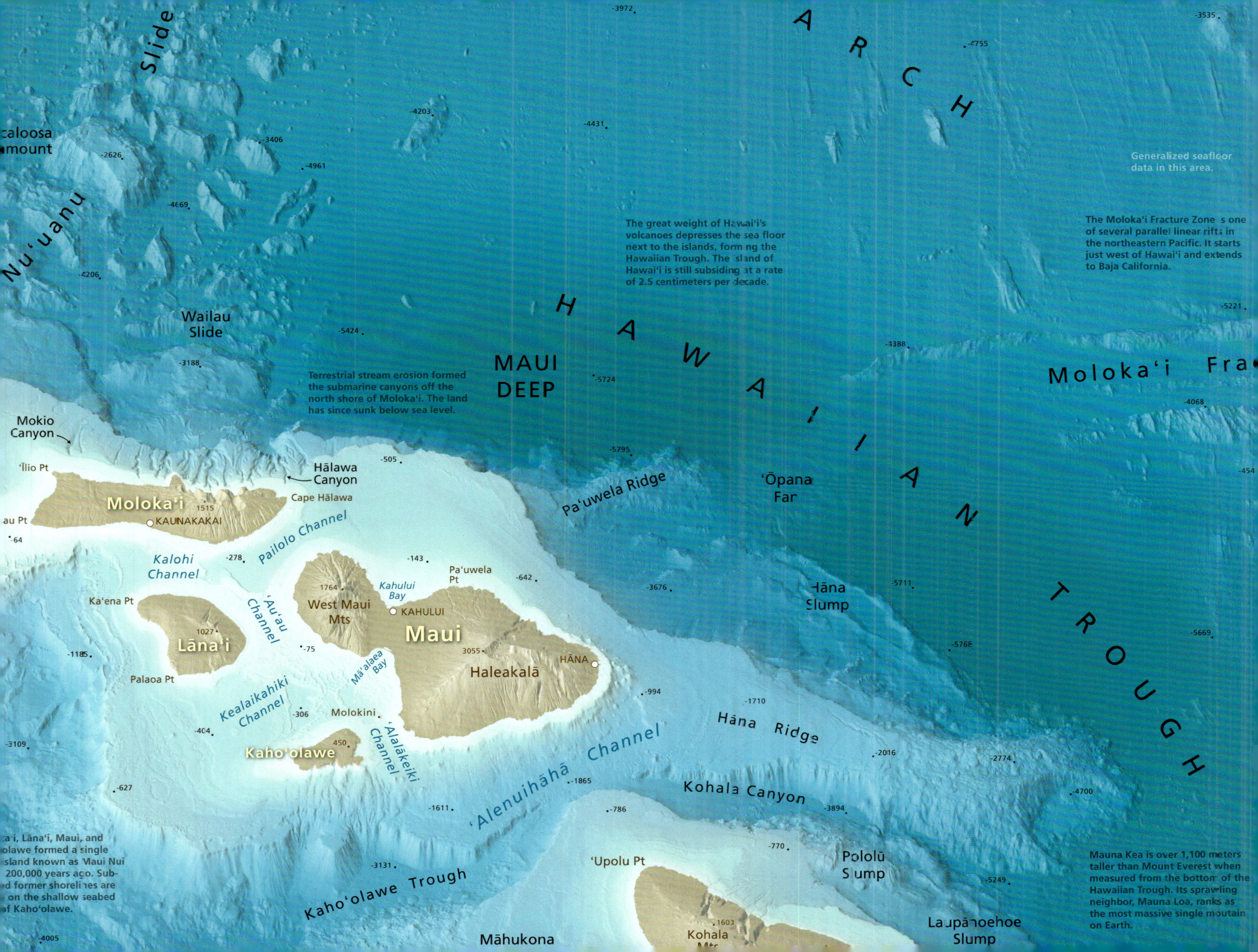

ARCH
HAWAIIAN TROUGH
Slide
Nu'uanu
caloosa
mount
Generalized seafloor data in this area.
The great weight of Hawai'i's volcanoes depresses the sea floor next to the islands, forming the Hawaiian Trough. The island of Hawai'i is still subsiding at a rate of 2.5 centimeters per decade.
The Moloka'i Fracture Zone is one of several parallel linear rifts in the northeastern Pacific. It starts just west of Hawai'i and extends to Baja California.
Wailau Slide
MAUI DEEP
Terrestrial stream erosion formed the submarine canyons off the north shore of Moloka'i. The land has since sunk below sea level.
Moloka'i Fra
Mokio Canyon
'Īlio Pt
Hālawa Canyon
Cape Hālawa
Pa'uwela Ridge
'Ōpana Fan
Moloka'i
KAUNAKAKAI
Pailolo Channel
Kalohi Channel
Pa'uwela Pt
Kahului Bay
KAHULUI
West Maui Mts
Hāna Slump
Ka'ena Pt
'Au'au Channel
Maui
Lāna'i
Haleakalā
HĀNA
Mā'alaea Bay
Palaoa Pt
Kealaikahiki Channel
Molokini
Kaho'olawe
'Alalākeiki Channel
Hāna Ridge
'Alenuihāhā Channel
Kohala Canyon
'Upolu Pt
Pololū Slump
Kaho'olawe Trough
Māhukona
Kohala Mts
Laupāhoehoe Slump
Mauna Kea is over 1,100 meters taller than Mount Everest when measured from the bottom of the Hawaiian Trough. Its sprawling neighbor, Mauna Loa, ranks as the most massive single moutain on Earth.

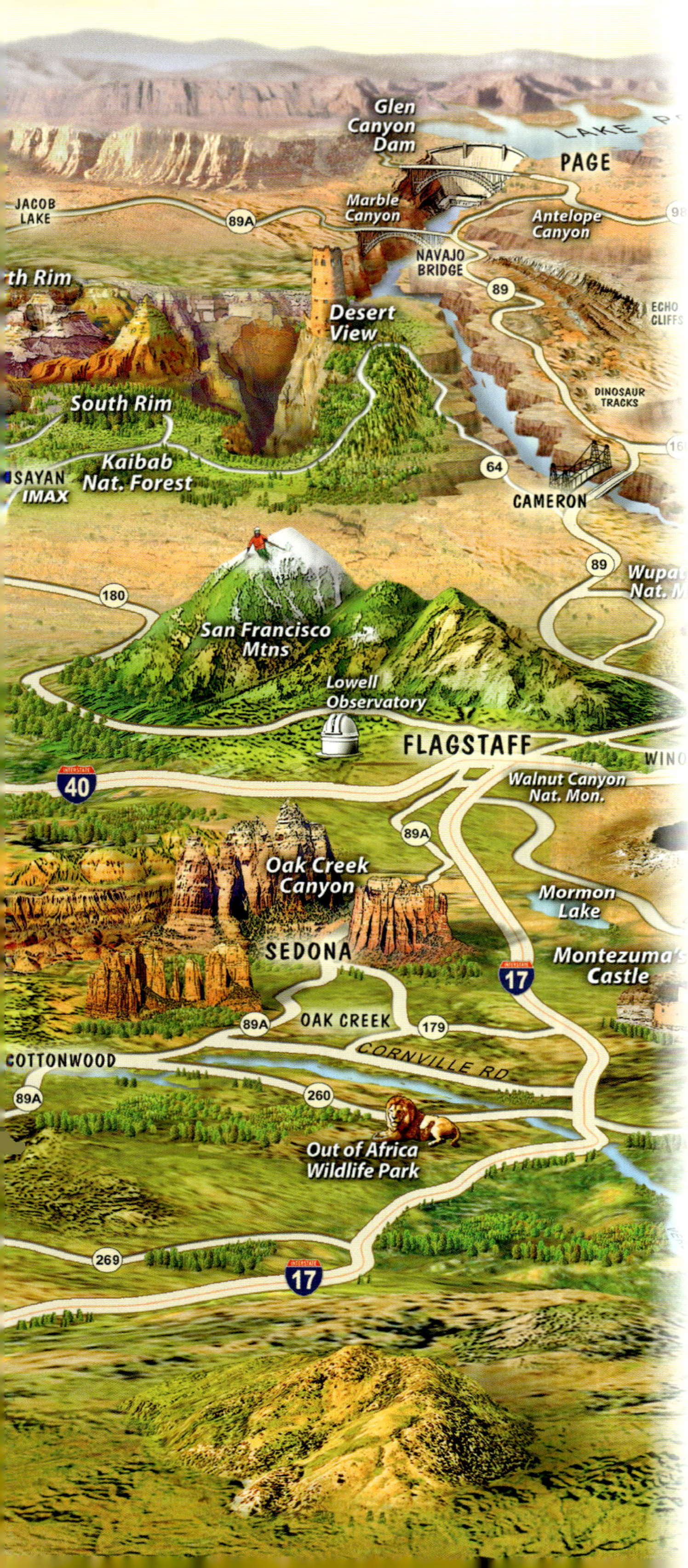

Northern Arizona »

Jean-Louis Rheault

Map originally created for Mapformation, LLC, client Randy Cooper 3D Maps

The mission of this map was to present a comprehensive, bird's eye view of Northern Arizona's key attractions and rich, colorful environment. Considering the vastness of the territory, any true 3D scale would make even the Grand Canyon a tiny squiggle. A pictorial mapping style was therefore chosen in order to emphasize the look and feel of individual landmarks as they are experienced.

The first step in creating this map involved pencil and paper. Building this sort of map is counterintuitive because it proceeds "upside down" from the way maps are usually constructed. Solo landmarks are sketched individually first. They are laid out with a play of interlocking scales in order to catch their best relational positions and familiar faces—similar to when a photographer moves people around for a group photo. Then, the road system and waterways are drawn in to connect the locations in a way that feels correct and organic. Lastly, the overall terrain landscape is sculpted to fit around the other elements. The resulting artwork is actually a set of multiple, ground-view vignettes, all sewn together to create the illusion of an overall oblique perspective.

Using the pencil layout as a template, the coloring and texturing is laid over like paint with many layers of Photoshop. Color is what holds everything together. For example, you can notice how the horizontal lines of ground texture create the impression of everything sitting on one terrain.

The art form of pictorial mapping is a centuries-old tradition that visualizes what cannot actually be shown by any vantage point, and it took me many years to break the deeply ingrained modern habit of constant scale. Whether executed with pen, pigment, or pixel, pictorial maps always help us understand places with the perspective of the imagination.

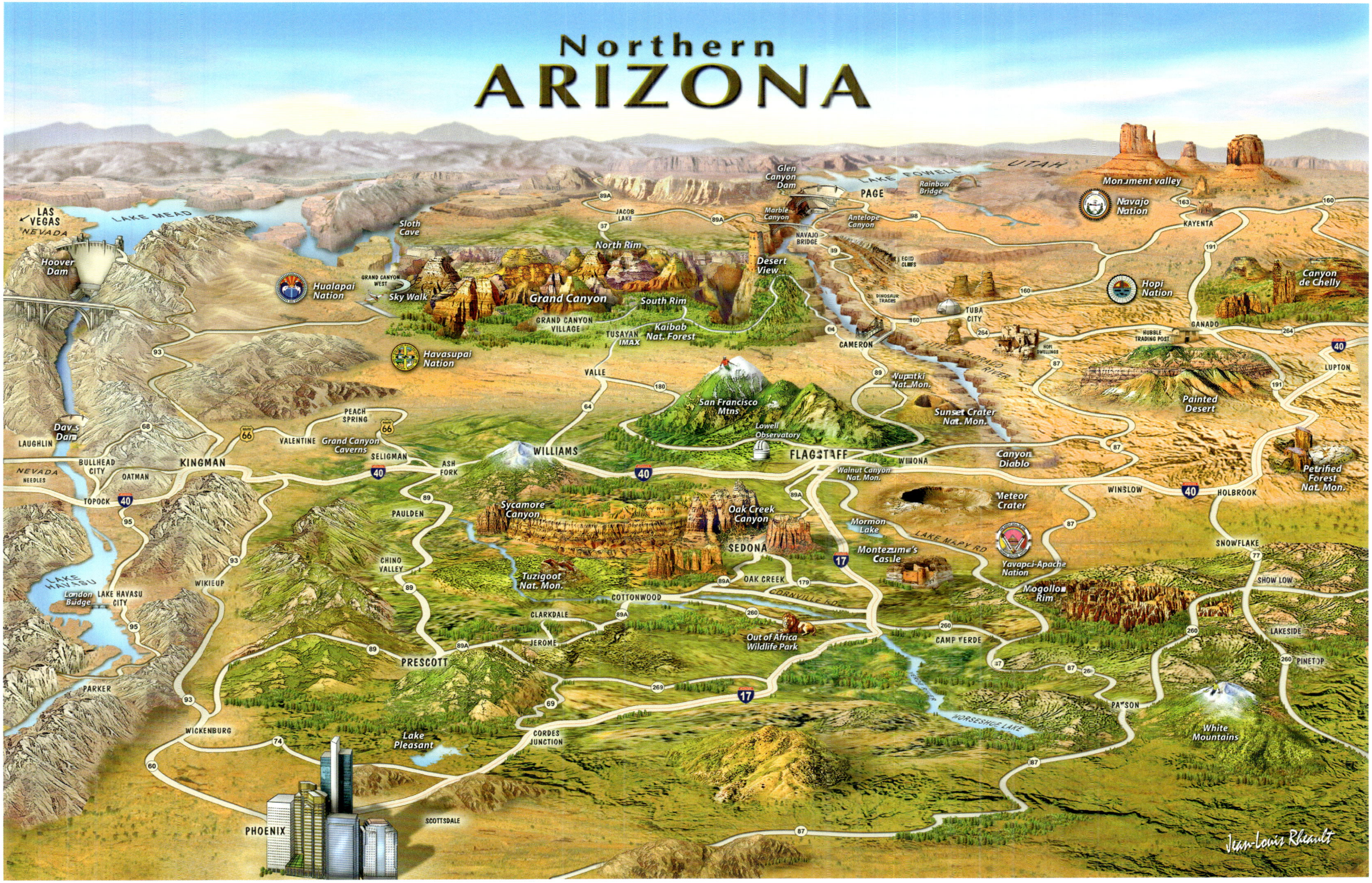
Northern
ARIZONA
LAS VEGAS
NEVADA
LAKE MEAD
Hoover Dam
Sloth Cave
Hualapai Nation
GRAND CANYON WEST
Sky Walk
Grand Canyon
North Rim
South Rim
GRAND CANYON VILLAGE
TUSAYAN IMAX
Kaibab Nat. Forest
Havasupai Nation
JACOB LAKE
Glen Canyon Dam
Marble Canyon
NAVAJO BRIDGE
Desert View
LAKE POWELL
PAGE
Antelope Canyon
Rainbow Bridge
UTAH
Monument valley
Navajo Nation
KAYENTA
Hopi Nation
Canyon de Chelly
TUBA CITY
CAMERON
GANADO
HUBBLE TRADING POST
LUPTON
Painted Desert
VALLE
San Francisco Mtns
Lowell Observatory
Wupatki Nat. Mon.
Sunset Crater Nat. Mon.
LITTLE COLORADO RIVER
FLAGSTAFF
WINONA
Canyon Diablo
Walnut Canyon Nat. Mon.
Meteor Crater
WINSLOW
HOLBROOK
Petrified Forest Nat. Mon.
WILLIAMS
ASH FORK
SELIGMAN
PEACH SPRING
VALENTINE
Grand Canyon Caverns
KINGMAN
LAUGHLIN
Davis Dam
BULLHEAD CITY
OATMAN
NEVADA
NEEDLES
TOPOCK
PAULDEN
Sycamore Canyon
Oak Creek Canyon
SEDONA
Mormon Lake
LAKE MARY RD
Montezuma's Castle
Yavapai-Apache Nation
SNOWFLAKE
SHOW LOW
Mogollon Rim
CHINO VALLEY
Tuzigoot Nat. Mon.
OAK CREEK
CORNVILLE RD
COTTONWOOD
CLARKDALE
JEROME
Out of Africa Wildlife Park
CAMP VERDE
LAKESIDE
PINETOP
WIKIEUP
LAKE HAVASU
London Bridge
LAKE HAVASU CITY
PRESCOTT
PARKER
PAYSON
HORSESHOE LAKE
White Mountains
WICKENBURG
Lake Pleasant
CORDES JUNCTION
PHOENIX
SCOTTSDALE
Jean-Louis Rheault

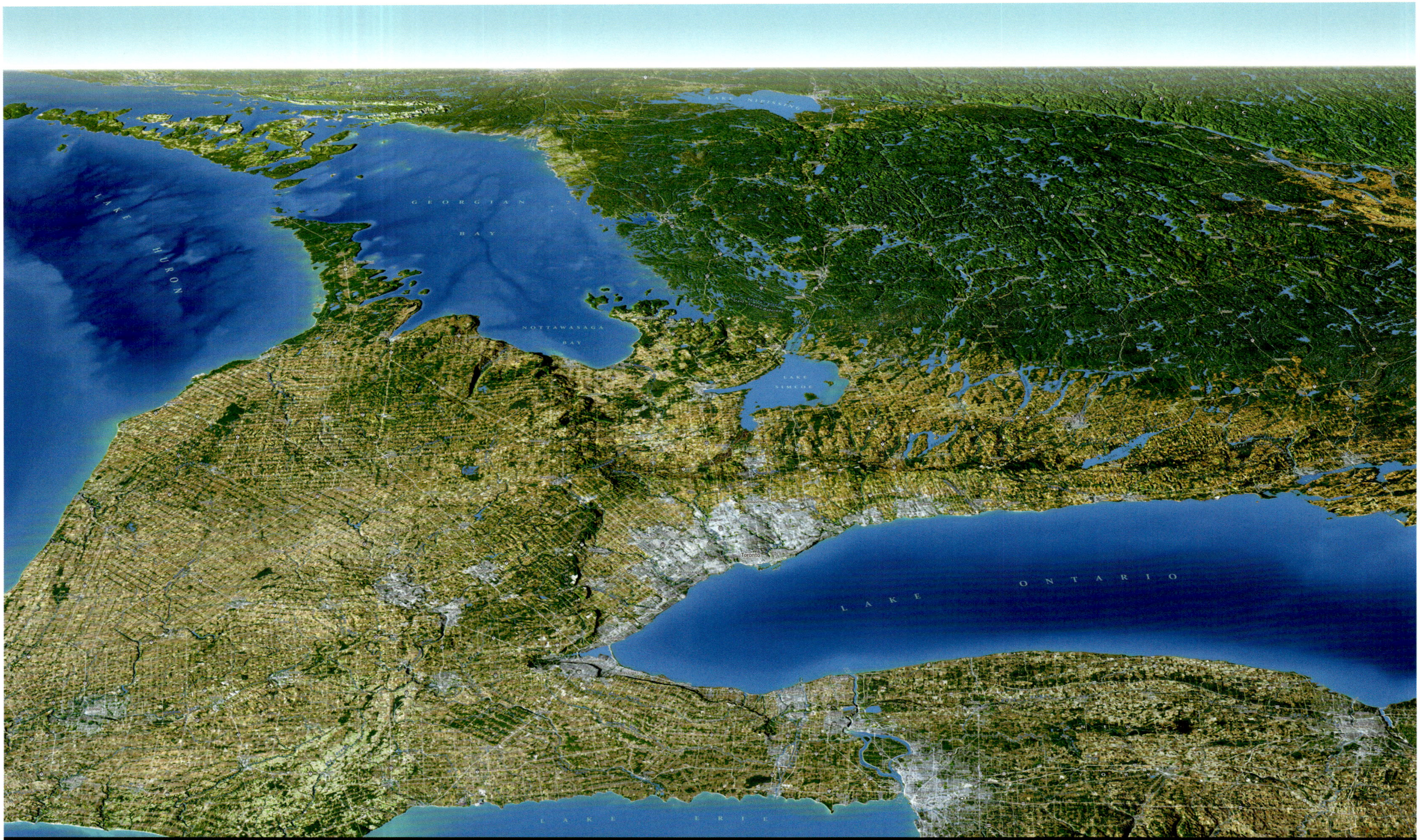

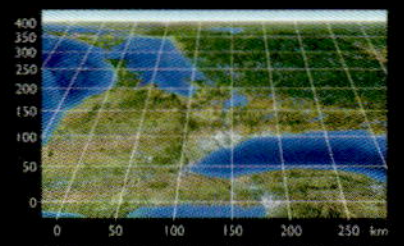

The grid overlay illustrates how scale varies across the map. Since the map plane is tilted, area in the foreground appears larger than in the background.

Shallow
Water Depth
Deep

Relative differences in bathymetry (measurement of water depth) are shown only for the Great Lakes

Urban area
Agriculture
Forest
Wetland
Bare rock
Pit/quarry

TORONTO METROPOLITAN REGION: THE BIG PICTURE

A PERSPECTIVE FROM 300 KILOMETRES UP

Today, with advances in technology such as Google Earth and Bing online maps, it is possible to zoom into intimate views of one's own backyard. But the bigger picture is often overlooked. It is only by zooming out that we can see the region as a whole, with all its interconnections. The Neptis Foundation, in collaboration with As the Crow Flies cARTography and the Cartography Office in the Department of Geography at the University of Toronto, created this unique view of the Toronto metropolitan region to help its residents and policymakers visualize those connections and better understand the region.

The Toronto region is home to about eight million people, and growing by one million people every decade. The businesses, industries, farms, and resources in the region produce nearly one-third of Canada's gross domestic product. The region is composed of a network of economically and functionally inter-related urban areas, a city of cities clustered around the City of Toronto, yet also an area with some of the best-quality agricultural land in Canada, thousands of hectares of forest, several unique landforms, and hundreds of fresh water lakes and rivers. The big picture shows how the region's metropolitan footprint borders three of the five Great Lakes. The dark green land cover in the northern quadrant of the map is the Canadian Shield, the largest area of exposed PreCambrian Rock on earth. It is home to Algonquin Park, Canada's first provincial park, established in 1893. The map also highlights prominent landforms, such as the Niagara Escarpment and the Oak Ridges Moraine, both of which are now part of the regional Greenbelt, which protects about 1.8 million acres of land.

The urban areas on the map stretch from the Niagara Peninsula in the south, to the cities of Kitchener and Waterloo in the west, to Collingwood, Barrie, and Orillia in the north and to Peterborough, Port Hope, and Cobourg in the east. The map includes dozens of separate municipalities, regional municipalities, and counties, all of which make decisions about urban form, environmental protection, transportation, water and waste management, and other matters. Yet the big picture reminds us that these entities do not act in isolation. A decision in one municipality can affect its neighbours, or even the region as a whole. The Toronto metropolitan area is a functioning system, in which every part contributes to the whole. The map was created by stitching together 16 satellite images (using Landsat 5 Thematic Mapper) and draping them over a digital and exaggerated model of the terrain. Unique characteristics of the landscape were highlighted by isolating and enhancing them in the imagery using thematic data on natural heritage lands, urbanized areas, roads, hydrography, and bathymetry (water depth). To further achieve a 3D perspective, the flat plane of the map is tilted. The perspective is the view you would get if you were nearly 300 kilometres in the air, hovering over Pennsylvania in the United States, looking north. Because of this unusual perspective, the scale varies across the map.

neptis THE NEPTIS FOUNDATION

Cartography and poster design by:
As the Crow Flies cARTography and The Cartography Office, Department of Geography, University of Toronto.

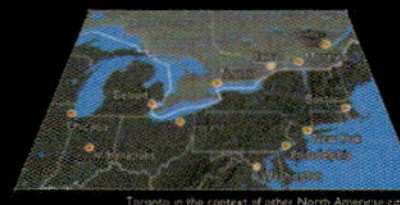

Toronto in the context of other North American cities

Data provided by: Neptis Foundation, University of Toronto, Ontario Ministry of Natural Resources © Her Majesty the Queen in Right of Canada, © Department of Natural Resources Canada, GeoBase®, United States Geological Survey, National Geophysical Data Center, Multi-Resolution Land Characteristics Consortium.

To order copies or obtain information on this poster, visit: www.neptis.org

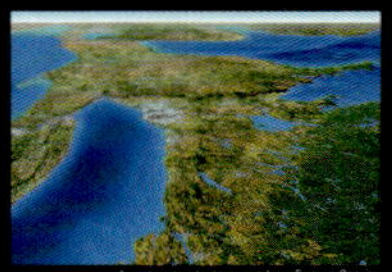

A unique view looking west from Eastern Ontario

Cartographic visualization has the power to take our experiential human-scale understanding of the world and apply it to a wider geographic extent.

« Toronto Metropolitan Region: the Big Picture — A Perspective from 300 Kilometres Up

Chris Brackley — As the Crow Flies cARTography; *The Neptis Foundation; The GIS and Cartography Office, Department of Geography, University of Toronto*

This map was designed to intuitively reveal patterns, relationships, and interconnections at a scale that draws new mental connections between the city, the farm, the quarry, and the cottage. It was generated by transforming satellite imagery from picture into map through a process of categorizing and coloring the landscape into profoundly familiar land cover types: agricultural, forested, and urban. We all know what it's like to experience these distinct environments, and by evoking each through the alchemy of color and texture, viewers can immediately place themselves on the map.

Add to this the birds-eye view perspective, where escarpments and moraines emerge and fall away, and where the land tapers towards the horizon, and this map truly achieves its ultimate aspiration, to be a *visualization*. Cartographic visualization has the power to take our experiential human-scale understanding of the world and apply it to a wider geographic extent—in this case the Toronto Metropolitan Region.

It is a region whose population is exploding, resulting in the transformation of prime agricultural land into paved urban spaces. Citizens and planners alike are well aware of this transformation, but rarely do they have the opportunity to see how their local reality relates to the big picture.

This map is intentionally designed to bridge these two scales. Labels are hidden within the map to be clear when standing close to it, but disappear when standing back. Local settings (e.g., big city parks, a gravel pit) are depicted in enough detail that viewers can place themselves onto the map, while at the same time the much larger area that surrounds them is depicted with a visual cohesion that allows the viewer to glimpse how it all connects.

The map shown here is really just a beginning. It is the baseline upon which policy and planning boundaries have been (and will continue to be) overlaid to reveal how the landscape will change over the coming years. Ideally, it will help guide thoughtful decision-making by citizens and governments in the years to come.

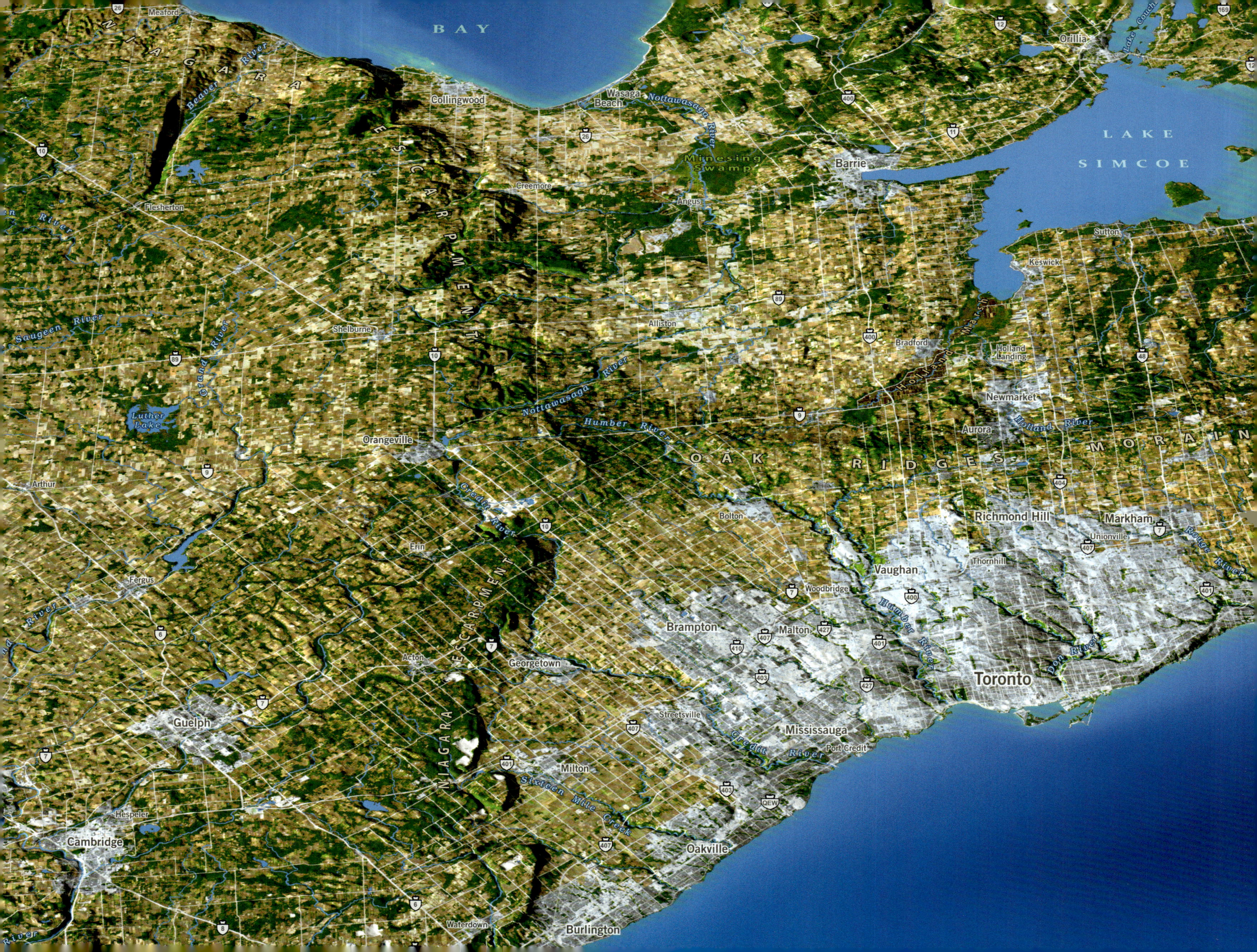

Meaford
BAY
NIAGARA ESCARPMENT
Beaver River
Collingwood
Wasaga Beach
Nottawasaga River
Orillia
LAKE SIMCOE
Minesing Swamp
Barrie
Creemore
Angus
Flesherton
River
Sutton
Keswick
Saugeen River
Grand River
Shelburne
Alliston
Bradford
Holland Marsh
Holland Landing
Newmarket
Aurora
Holland River
Luther Lake
Orangeville
Humber River
OAK RIDGES MORAINE
Arthur
Credit River
Bolton
Richmond Hill
Markham
Unionville
Rouge River
Erin
Fergus
Vaughan
Thornhill
Woodbridge
River
Brampton
Malton
Humber River
Acton
Georgetown
Don River
Toronto
Guelph
Streetsville
Mississauga
Port Credit
Credit River
NIAGARA ESCARPMENT
Milton
Sixteen Mile Creek
Hespeler
Cambridge
Oakville
Waterdown
Burlington
River

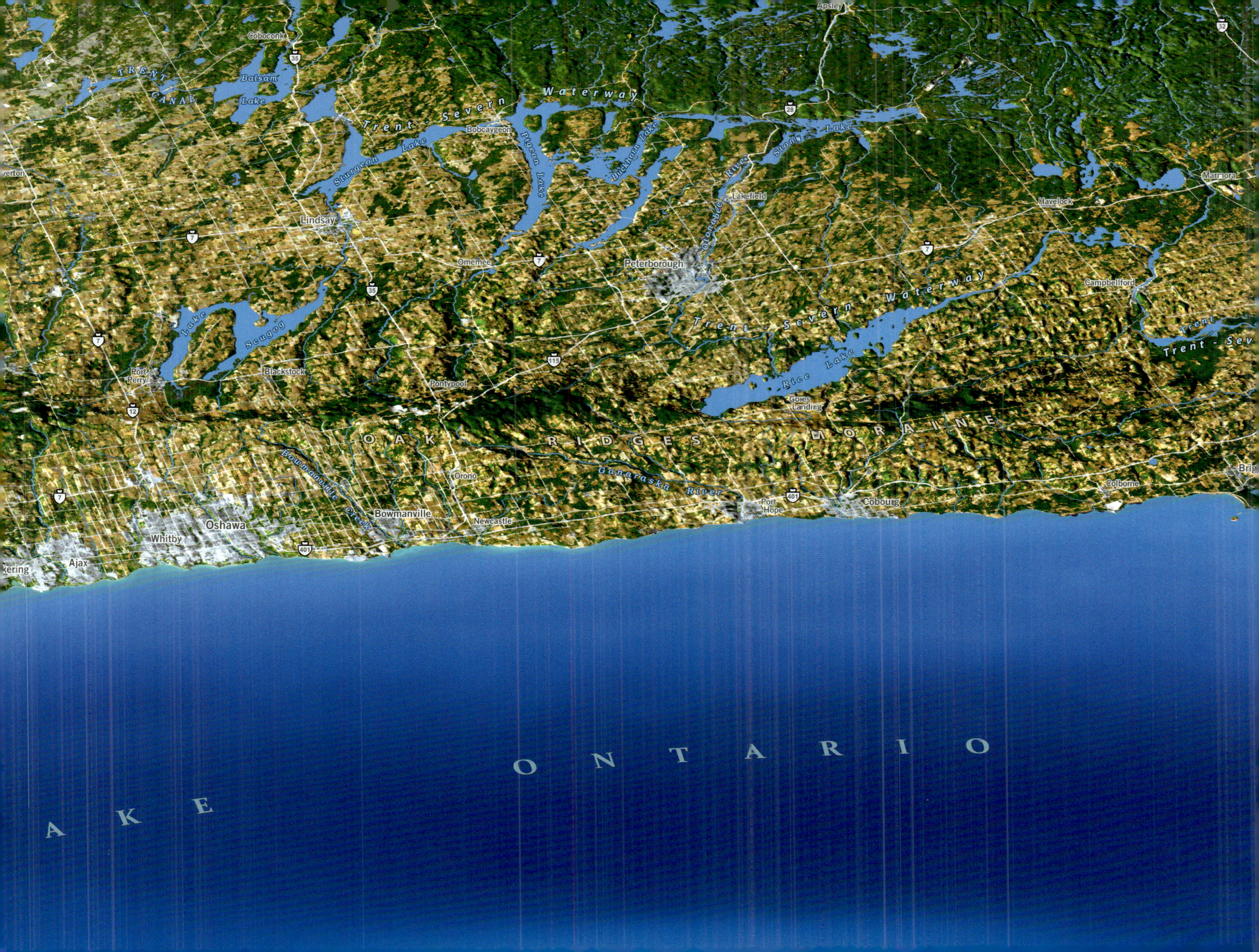
Apsley
Coboconk
35
Balsam Lake
Trent Canal
Trent - Severn Waterway
Bobcaygeon
28
Sturgeon Lake
Pigeon Lake
Buckhorn Lake
Stony Lake
Otonabee River
Lakefield
Marmora
Havelock
Lindsay
7
Omemee
Peterborough
Campbellford
Trent - Severn Waterway
Rice Lake
Lake Scugog
Port Perry
Blackstock
115
Pontypool
12
Gores Landing
OAK RIDGES MORAINE
Bowmanville Creek
Ganaraska River
Orono
Bowmanville
Newcastle
Port Hope
401
Cobourg
Colborne
Oshawa
Whitby
Ajax
AKE ONTARIO

» *Although technically unnecessary, we incorporated a number of stylistic novelties to augment the overall design and effect of this map.*

Tea Horse Road »

Martin Gamache, Marguerite B. Hunsiker, Fernando G. Baptista, Mike Reagan, Elaine Bradley, Elbie Bentley — National Geographic Magazine

Our design aesthetic for this map was informed by a map published in 1919 by the Royal Geographical Society that depicted Oliver Coates' travels in this part of Sichuan and Tibet. With that aesthetic in mind, we hired Michael Reagan, a cartographer specializing in watercolor maps, to paint a map background that could be combined with thematic elements. We also extensively researched the caravan routes, the value and quantity of tea being traded in these areas, and the modes of transportation. A lack of accurate data ultimately dictated a less thematic mapping approach.

Reagan's map art was scanned and integrated with our cartographic database. We used Photoshop to clean up and capture the textures and pictographic elements of his artwork, being careful to retain Reagan's artistic contribution while maintaining our accuracy requirements. Interestingly, most modern roads in Tibet, built up over the last 20 years of Chinese occupation, follow original trade routes. This fact allowed us to source more modern road maps and databases, such as OpenStreetMap, when mapping the major trade routes.

Although technically unnecessary, we incorporated a number of stylistic novelties to augment the overall design and effect of this map. These include breaking the neatline in many places in order to have map elements (roads, water, boundaries) bleed off the page, and pictorial elements, such as painted trees (used to indicate tea-growing areas) and painted mountains (to indicate key areas of relief). We broke with our typical sans serif house fonts for notes, credits, and scale bars, using inked lines on the scale bars to match the period style.

We employed standard techniques for historical map re-creation such as scanning stained, creased, and aged paper for use as a map background. We also filtered rasterized linework and added noise to reproduce the look and feel of a historic map. At times we have seen such techniques applied with too strong a hand, so we employed a subtler touch to hint at, rather than scream, our intentions (e.g., we opted to omit a gaudy, overly elaborate, nautical-themed north arrow).

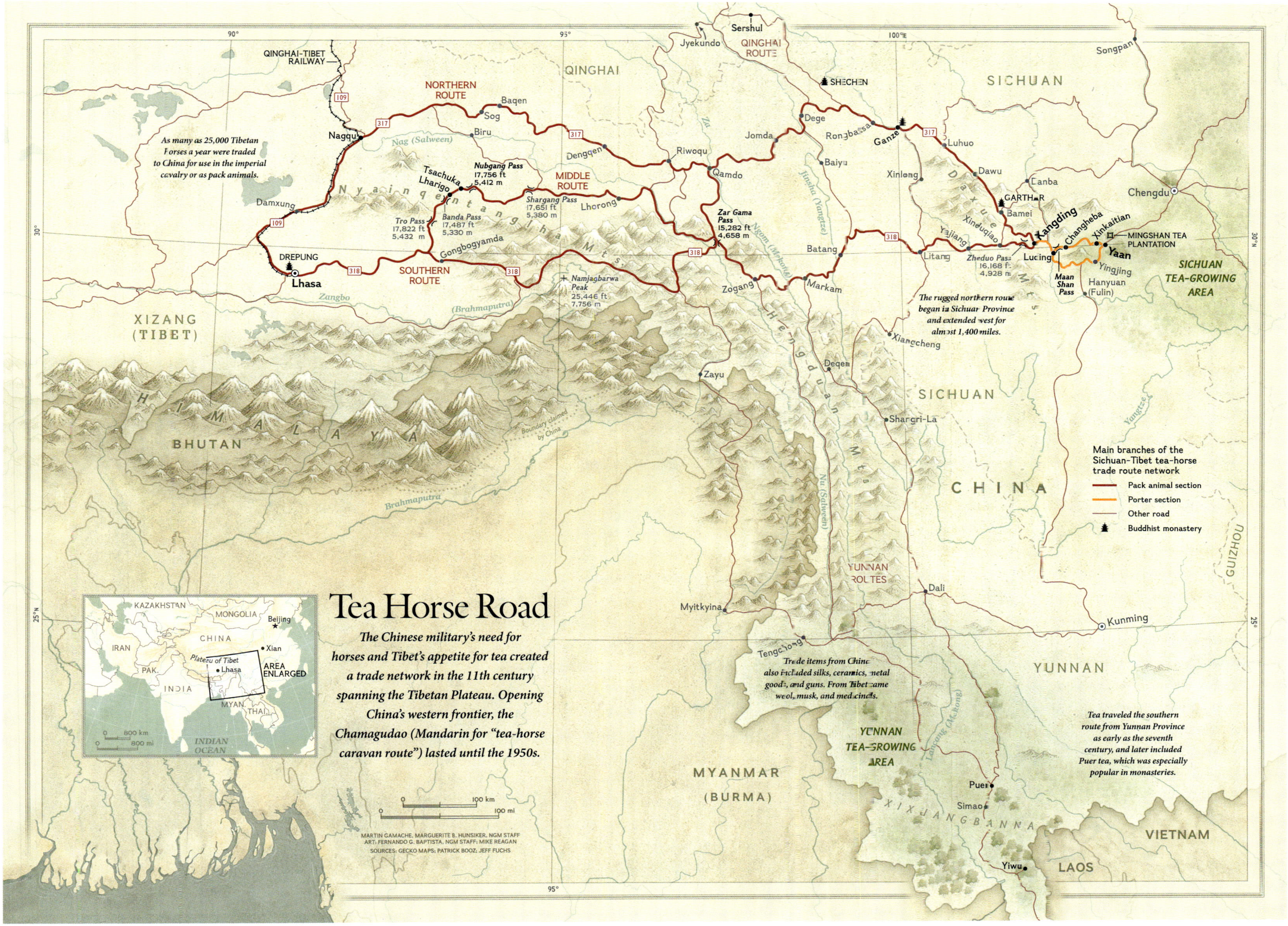

Tea Horse Road
The Chinese military's need for horses and Tibet's appetite for tea created a trade network in the 11th century spanning the Tibetan Plateau. Opening China's western frontier, the Chamagudao (Mandarin for "tea-horse caravan route") lasted until the 1950s.
As many as 25,000 Tibetan horses a year were traded to China for use in the imperial cavalry or as pack animals.
The rugged northern route began in Sichuan Province and extended west for almost 1,400 miles.
Trade items from China also included silks, ceramics, metal goods, and guns. From Tibet came wool, musk, and medicines.
Tea traveled the southern route from Yunnan Province as early as the seventh century, and later included Puer tea, which was especially popular in monasteries.
Main branches of the Sichuan-Tibet tea-horse trade route network
Pack animal section
Porter section
Other road
Buddhist monastery
NORTHERN ROUTE
MIDDLE ROUTE
SOUTHERN ROUTE
QINGHAI ROUTE
YUNNAN ROUTES
QINGHAI-TIBET RAILWAY
QINGHAI
SICHUAN
XIZANG (TIBET)
CHINA
YUNNAN
BHUTAN
MYANMAR (BURMA)
LAOS
VIETNAM
GUIZHOU
SICHUAN TEA-GROWING AREA
YUNNAN TEA-GROWING AREA
MINGSHAN TEA PLANTATION
Lhasa
Nagqu
Damxung
Qamdo
Kangding
Yaan
Chengdu
Kunming
Dali
Puer
Simao
Yiwu
Shangri-La
Nubgang Pass 17,756 ft 5,412 m
Tro Pass 17,822 ft 5,432 m
Banda Pass 17,487 ft 5,330 m
Shargang Pass 17,651 ft 5,380 m
Zar Gama Pass 15,282 ft 4,658 m
Zheduo Pass 16,168 ft 4,928 m
Namjagbarwa Peak 25,446 ft 7,756 m
Kazakhstan
Mongolia
Beijing
Xian
Plateau of Tibet
AREA ENLARGED
Indian Ocean
MARTIN GAMACHE, MARGUERITE B. HUNSIKER, NGM STAFF
ART: FERNANDO G. BAPTISTA, NGM STAFF; MIKE REAGAN
SOURCES: GECKO MAPS; PATRICK BOOZ; JEFF FUCHS

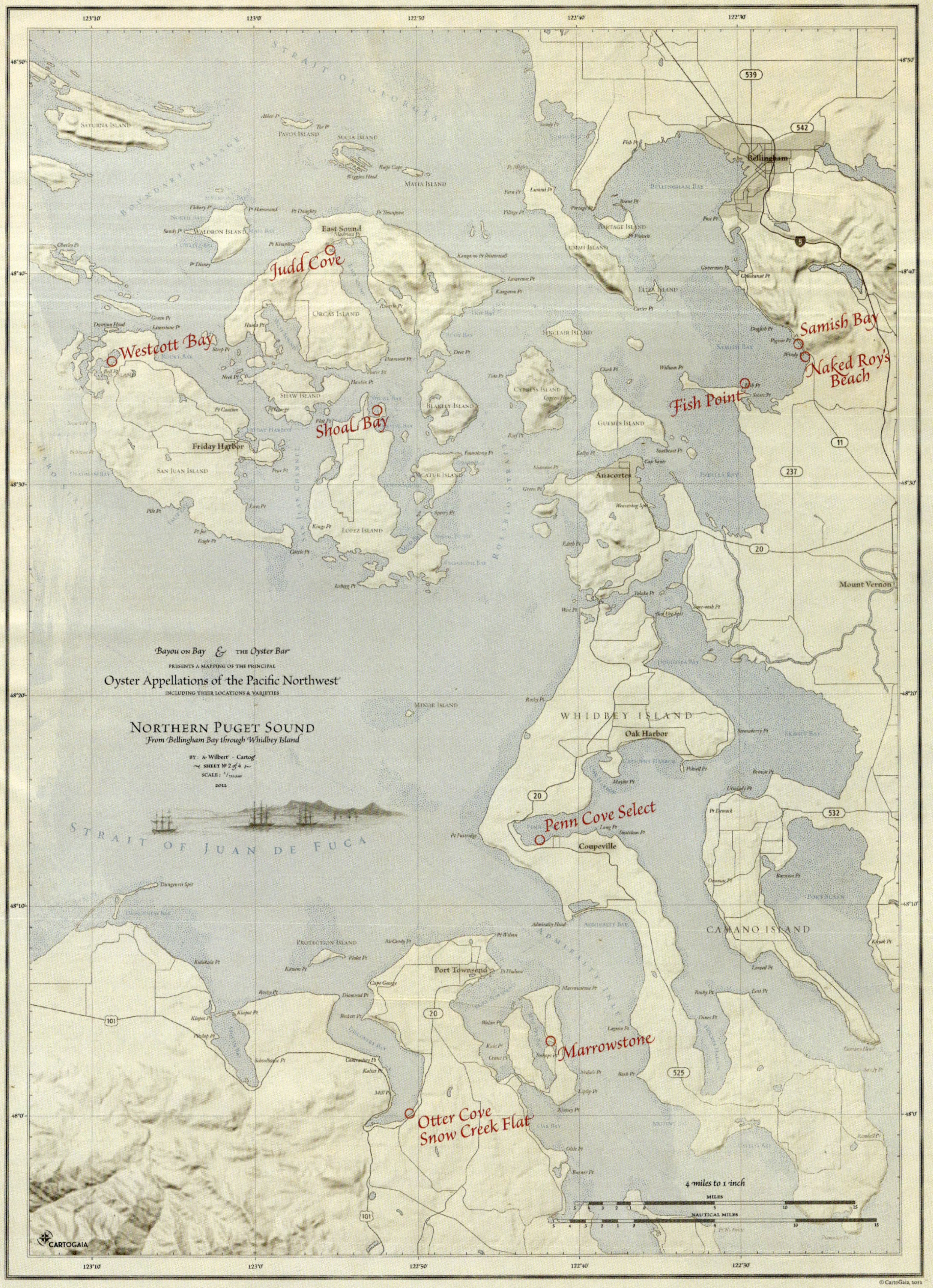
Bayou on Bay & the Oyster Bar
presents a mapping of the principal
Oyster Appellations of the Pacific Northwest
including their locations & varieties
Northern Puget Sound
From Bellingham Bay through Whidbey Island
By: A. Wilbert · Cartog
Sheet № 2 of 4
2012
Judd Cove
Westcott Bay
Shoal Bay
Fish Point
Samish Bay
Naked Roy's Beach
Penn Cove Select
Marrowstone
Otter Cove
Snow Creek Flat
Strait of Georgia
Strait of Juan de Fuca
Boundary Passage
Rosario Strait
San Juan Channel
Admiralty Inlet
Saturna Island
Patos Island
Sucia Island
Matia Island
Waldron Island
East Sound
Orcas Island
Shaw Island
Friday Harbor
San Juan Island
Lopez Island
Blakely Island
Decatur Island
Cypress Island
Sinclair Island
Guemes Island
Lummi Island
Portage Island
Eliza Island
Anacortes
Bellingham
Bellingham Bay
Samish Bay
Mount Vernon
Whidbey Island
Oak Harbor
Coupeville
Camano Island
Minor Island
Protection Island
Port Townsend
Dungeness Spit
4 miles to 1 inch
Miles
Nautical Miles
Cartogaia
©CartoGaia, 2012

« Oyster Appellations of the Pacific Northwest, Sheet 2 of 4: Northern Puget Sound

Adam Wilbert — CartoGaia

Long before I was interested in the profession of mapping, I was interested in the art of printing. The study of putting ink on paper, and how that process has evolved over the past 500 years, is absolutely fascinating to me. The art of etching copper and pressing it into a sheet; the photo-mechanics of lithography, letterpress, and screen printing; the choice of registering colored plates on press, or hand washing a sheet with watercolor hues; the chemistry of ink and acids and lignin and age . . . all these contribute to the tactile and humanized art of printing that I strive to incorporate into my work. In part, this map is a reaction to how cold and precise maps in our digital world have become. In my mind, the warmth of decaying old paper and rusting iron ink restores a bit of romance to the journey through a map.

This map is one in a series that will hang in a local restaurant. The cartographic style is modeled after the charts from George Vancouver's visit to Puget Sound, published in 1798, and the U.S. Coast Survey, published in 1853. The growing and harvest locations for each oyster variety are marked as if they were annotated by hand long after the initial publication of the map.

Various masks were used to create the aged effects that were modeled after real-world processes. Copperplate etchings are prone to wear during use and cleanup. Often this results in the erosion of fine lines in subsequent printings. Older maps and charts were often hand colored with watercolor washes after the initial printing of the line work. This exposes irregularities in the texture of the paper stock and leaves brush strokes in large areas of color. It is these subtle quirks in printmaking that give a printed piece character and personality unmatched by digitally created solid fills and impossibly perfect lines.

In part, this map is a reaction to how cold and precise maps in our digital world have become.

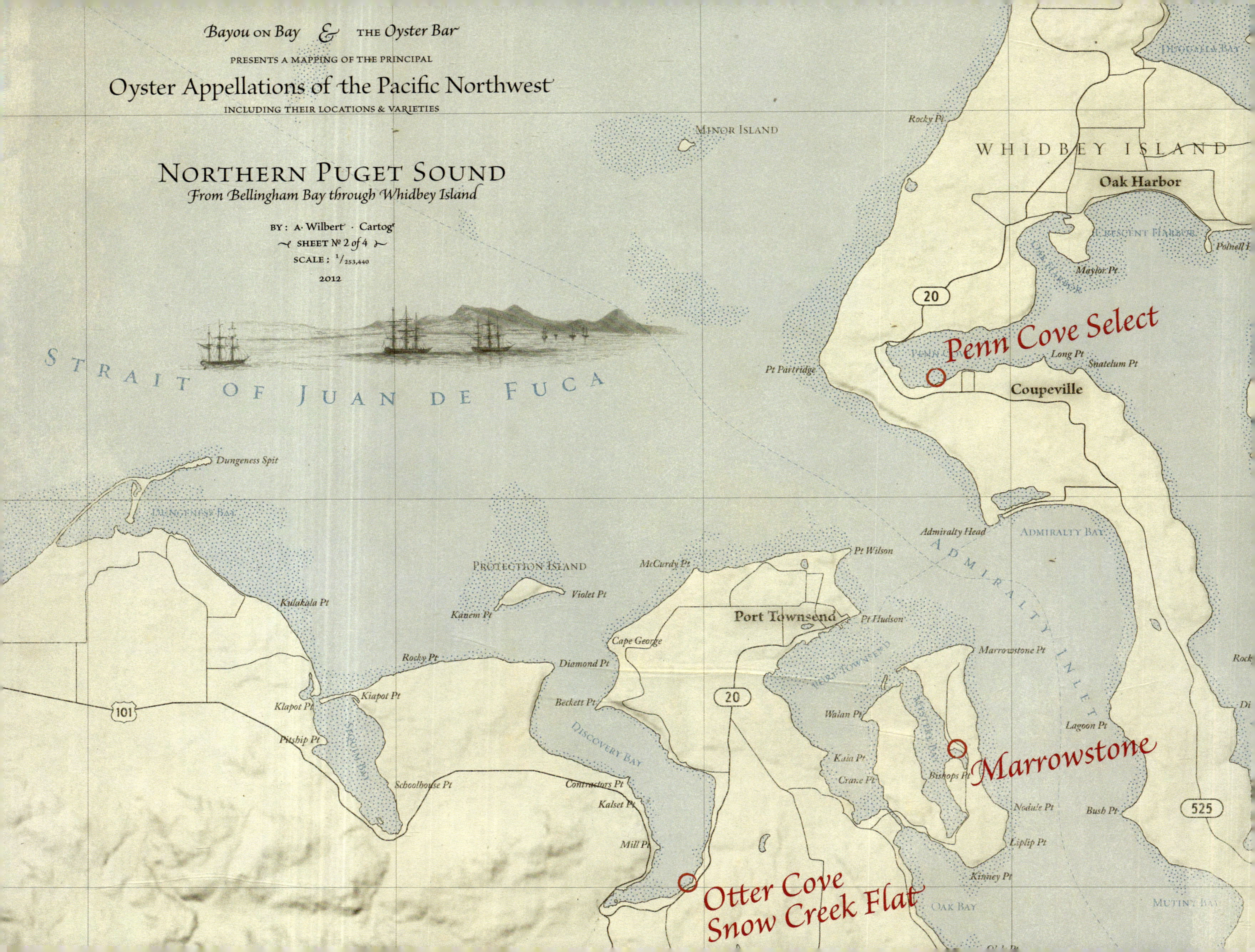
Bayou on Bay & the Oyster Bar
presents a mapping of the principal
Oyster Appellations of the Pacific Northwest
including their locations & varieties
Northern Puget Sound
From Bellingham Bay through Whidbey Island
By: A. Wilbert · Cartogr
Sheet № 2 of 4
Scale: 1/253,440
2012
Strait of Juan de Fuca
Minor Island
Rocky Pt
Whidbey Island
Oak Harbor
Crescent Harbor
Polnell Pt
Maylor Pt
Oak Harbor
20
Penn Cove Select
Long Pt
Snatelum Pt
Pt Partridge
Coupeville
Dungeness Spit
Dungeness Bay
Admiralty Head
Admiralty Bay
Admiralty Inlet
Pt Wilson
Protection Island
McCurdy Pt
Violet Pt
Kanem Pt
Kulakala Pt
Port Townsend
Pt Hudson
Cape George
Marrowstone Pt
Port Townsend
Rocky Pt
Diamond Pt
Kiapot Pt
Klapot Pt
101
20
Beckett Pt
Walan Pt
Discovery Bay
Sequim Bay
Pitship Pt
Lagoon Pt
Marrowstone
Kala Pt
Crane Pt
Bishops Pt
Schoolhouse Pt
Contractors Pt
Kalset Pt
Nodule Pt
Bush Pt
525
Mill Pt
Liplip Pt
Kinney Pt
Otter Cove
Snow Creek Flat
Oak Bay
Mutiny Bay

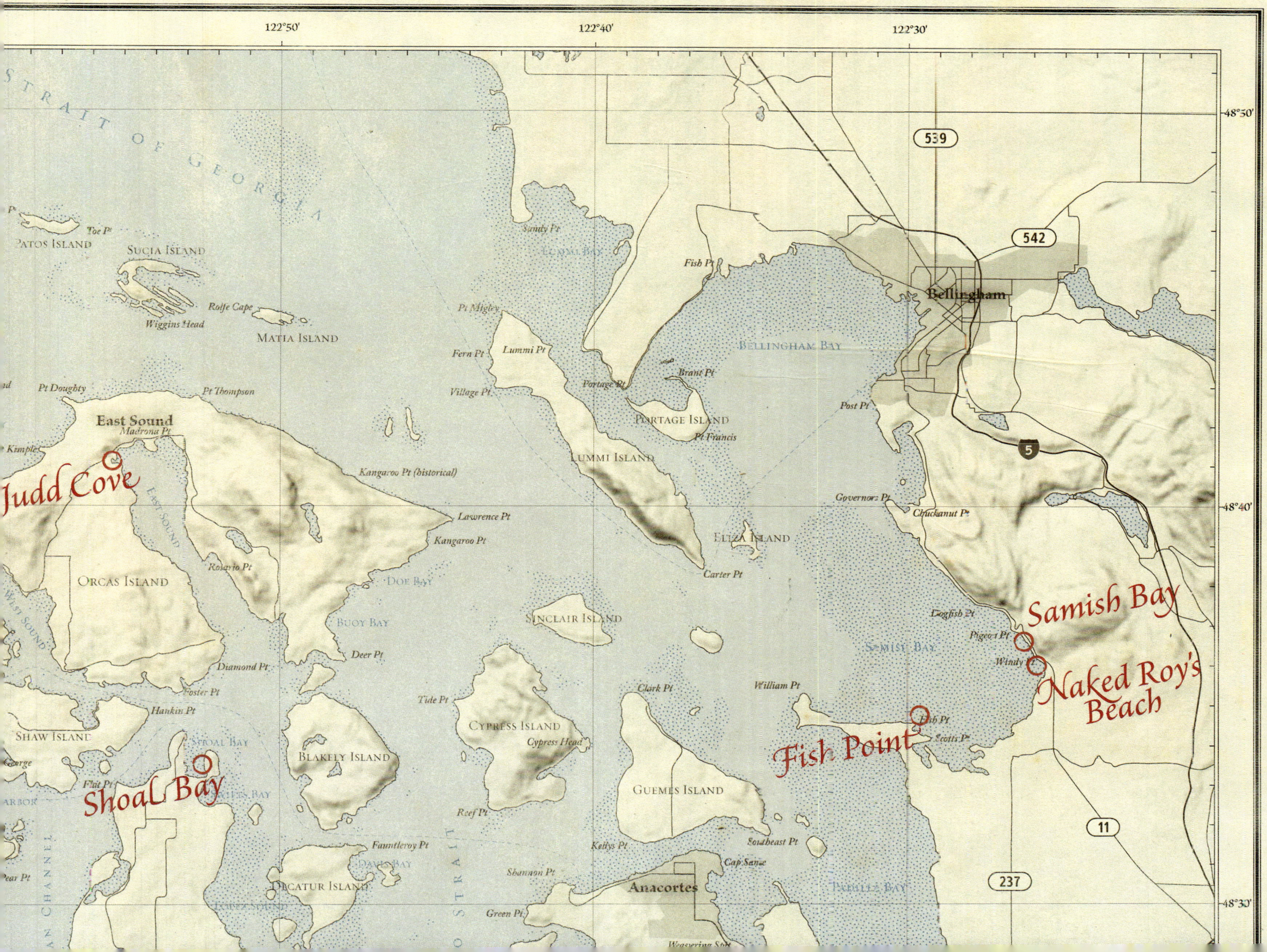
122°50'
122°40'
122°30'
48°50'
48°40'
48°30'
STRAIT OF GEORGIA
PATOS ISLAND
Toe Pt
SUCIA ISLAND
Rolfe Cape
Wiggins Head
MATIA ISLAND
Pt Doughty
Pt Thompson
East Sound
Madrona Pt
Kimple
Judd Cove
EAST SOUND
ORCAS ISLAND
WEST SOUND
Kangaroo Pt (historical)
Lawrence Pt
Kangaroo Pt
Rosario Pt
DOE BAY
BUOY BAY
Deer Pt
Diamond Pt
Foster Pt
Hankin Pt
SHAW ISLAND
SHOAL BAY
Flat Pt
Shoal Bay
BLAKELY ISLAND
DECATUR ISLAND
Fauntleroy Pt
DAVIS BAY
LOPEZ SOUND
SINCLAIR ISLAND
Tide Pt
CYPRESS ISLAND
Cypress Head
Reef Pt
Clark Pt
GUEMES ISLAND
Kellys Pt
Shannon Pt
Green Pt
Anacortes
Cap Sante
Southeast Pt
William Pt
Fish Pt
Scotts Pt
Fish Point
SAMISH BAY
Samish Bay
Pigeon Pt
Windy
Naked Roy's Beach
Dogfish Pt
Chuckanut Pt
Governors Pt
ELIZA ISLAND
Carter Pt
LUMMI ISLAND
PORTAGE ISLAND
Pt Francis
Brant Pt
Portage Pt
Post Pt
BELLINGHAM BAY
Bellingham
Fish Pt
Sandy Pt
LUMMI BAY
Pt Migley
Lummi Pt
Fern Pt
Village Pt
PADILLA BAY
539
542
5
11
237

Atlas of Florida's Natural Heritage: Biodiversity, Landscapes, Stewardship, and Opportunities »

Tanya M. A. Buckingham — University of Wisconsin Cartography Lab; *Lou Cross* — Florida State University–Institute of Science and Public Affairs

The maps and layout for the special Notable Natural Areas section in the *Atlas of Florida's Natural Heritage* shifted the emphasis of the atlas to place, and tied together the individual elements of natural heritage to highlight outstanding natural places in Florida. The spreads in this section depart from the detailed, scientific presentation used elsewhere in the atlas, and evoke instead a field notebook style, wherein the maps appear to be produced by a hand-drawn, loose, pen/ink, and/or watercolor medium.

Initial design concepts of the 12 maps in the series attempted to soften the discrete GIS polygon data through the use of blurred/softened edges and semi-transparent fills. However, those treatments were quickly deemed too heavy and literal, and therefore a style was adopted to mimic a field researcher, quickly highlighting the area with a thick yellow marker on the map that was sketched in their field notebook. The highlight also serves as the legend, and ties the title of the natural area to the feature on the map.

While the design goal was for the overall piece to be "sketchy" in feel, accuracy and detail were also required. The map data (e.g., roads, county boundaries and labels, and named conservation lands) needed to be included for reference. Given these parameters, the overall feel of the map was to be hand-painted, and hand-lettered—but legible and detailed. The look of the map was created by softening the edges of the fills, selecting certain brush strokes, adding textures and blurs, and running a few filters. Linework was set to overprint, and tapered where possible.

An alternative design approach that was considered included the map cut and taped to the page—as if it had been torn from a travel atlas, as you see with the photos and charts. In the end, however, this added heaviness to the page, and did not contribute to the artistic element that the maps add to the final overall piece. Ultimately, trying to balance the detail of the maps with a hand-drawn feeling was challenging, but rewarding.

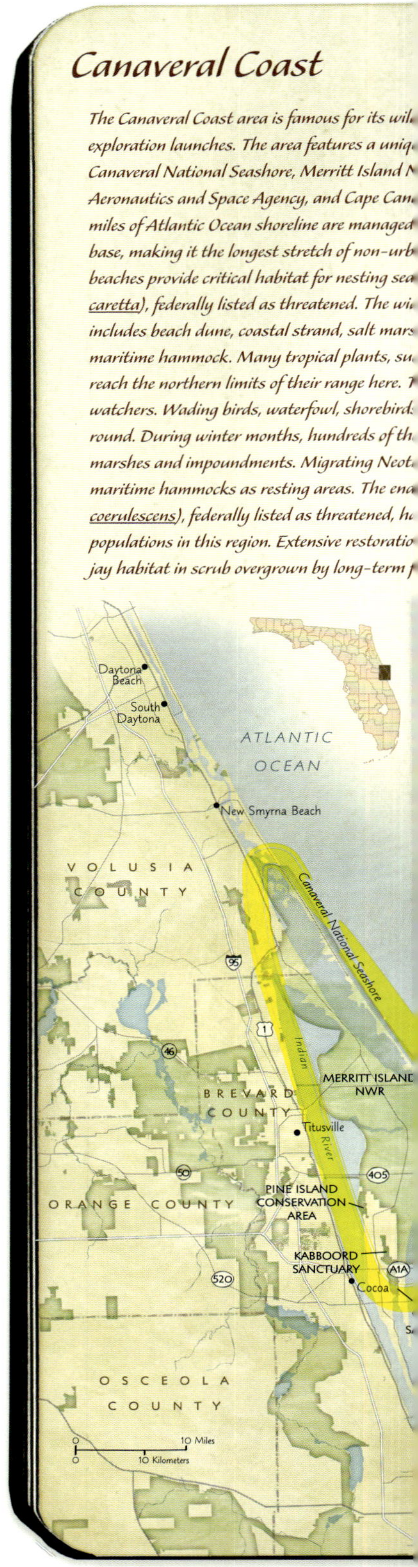

…da scrub-jay (Aphelocoma coerulescens) →

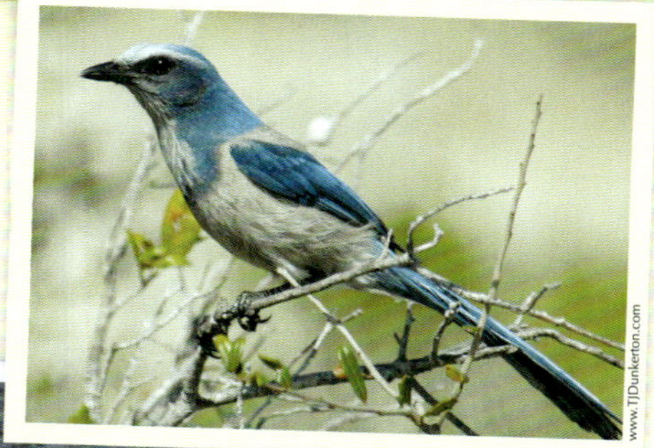

…ed beaches, and spac… partnership between …fe Refuge, National … Station. About 40 …ashore and military … beach in Florida. The …ally loggerhead (Caretta … …tural communities …bby flatwoods, and … and gumbo limbo, …ime destination for bird …d raptors forage year …ratory birds use the …ds rely on the coastal …rub-jay (Aphelocoma …gest remaining …going to improve scrub-

Roseate spoonbills (Platalea ajaja) ↓

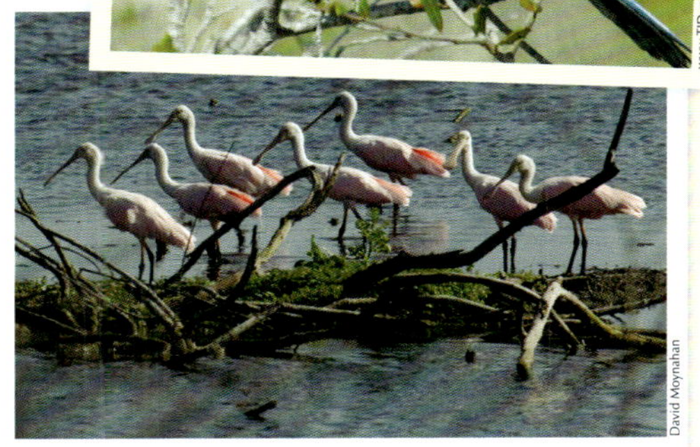

…S OF BIODIVERSITY

…ities

…ock

…anulicola	sand-dune spurge
…ii	Simpson's prickly apple
…sa var.	Atlantic Coast Florida lantana
…stissima	coastal hoary-pea
…rulescens	Florida scrub-jay
	loggerhead
…s	green turtle
…riacea	leatherback
…uperi	eastern indigo snake
…ana	wood stork

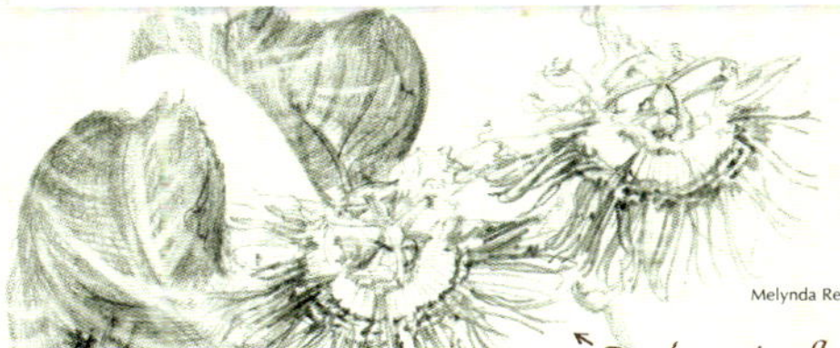

Purple passion-flower (Passiflora incarnata), a common vine

…al Coast

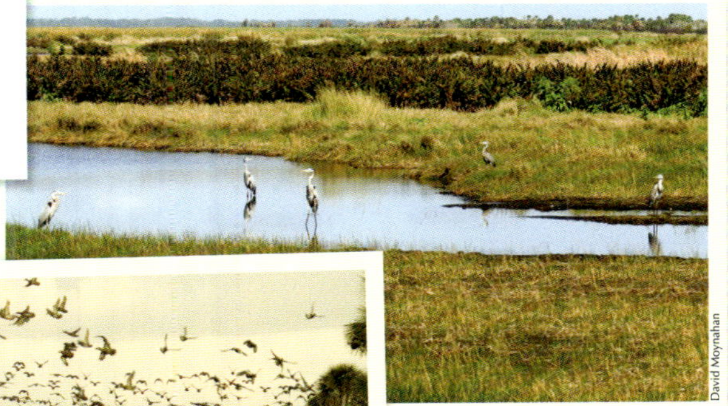

↑ *Great blue herons (Ardea herodias) foraging in tidal wetlands*

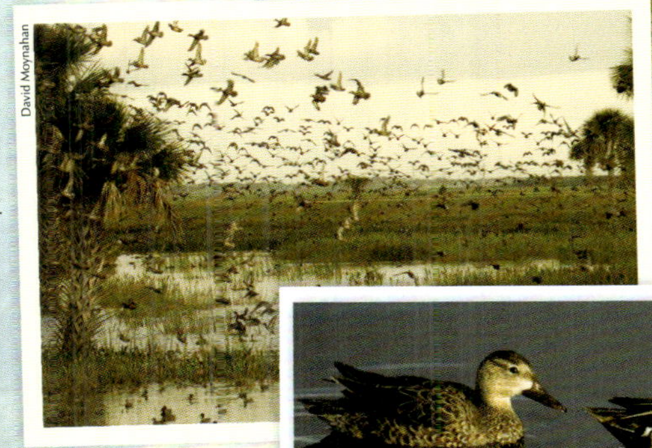

↑ *Waterfowl are often abundant in winter months*

…LANTIC …OCEAN

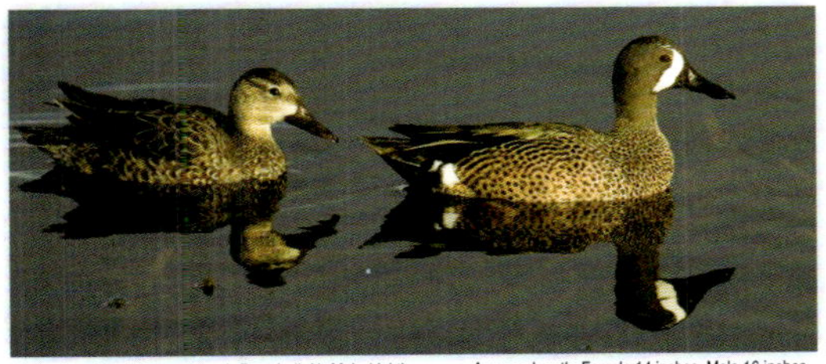

Blue-winged teal (*Anas discors*), Female (left), Male (right)

Average length: Female 14 inches, Male 16 inches
Average weight: Female 0.8 lbs., Male 1.0 lbs.

Lake Wales Ridge

The Lake Wales Ridge is a series of ancient sand dunes deposited about 650,000 years ago when much of the Florida peninsula was under water. This long and narrow former chain of islands was separated from the mainland, causing its plants and animals to evolve in isolation. As a result, the ridge now contains one of the highest concentrations of rare plants and animals in the United States. The importance of the ridge to rare plant conservation in Florida—and globally—cannot be overstated. Twenty-two (39%) of Florida's 56 federally listed plants are found here. The ecological importance of the area is reflected in the creation of the Lake Wales Ridge National Wildlife Refuge, the nation's first wildlife refuge created for the recovery of endangered and threatened plants. More than 85% of the original 80,000 acres of upland habitats on the Lake Wales Ridge has been lost to development. Remnants of intact habitat are found scattered up and down the approximately 115-mile-long and up to 10-mile-wide area. Each has its own distinctive suite of species, demonstrating the need to conserve areas throughout the length of the ridge. The Carter Creek Unit of the Lake Wales Ridge Wildlife and Environmental Area, Lake Wales Ridge State Forest, and TNC's Tiger Creek Preserve are some of the best remaining examples of the original landscape. All offer exceptional opportunities to see many rare plants and animals, such as Florida scrub-jay (Aphelocoma coerulescens), dependent on these unique habitats.

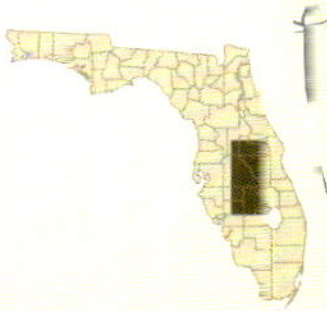

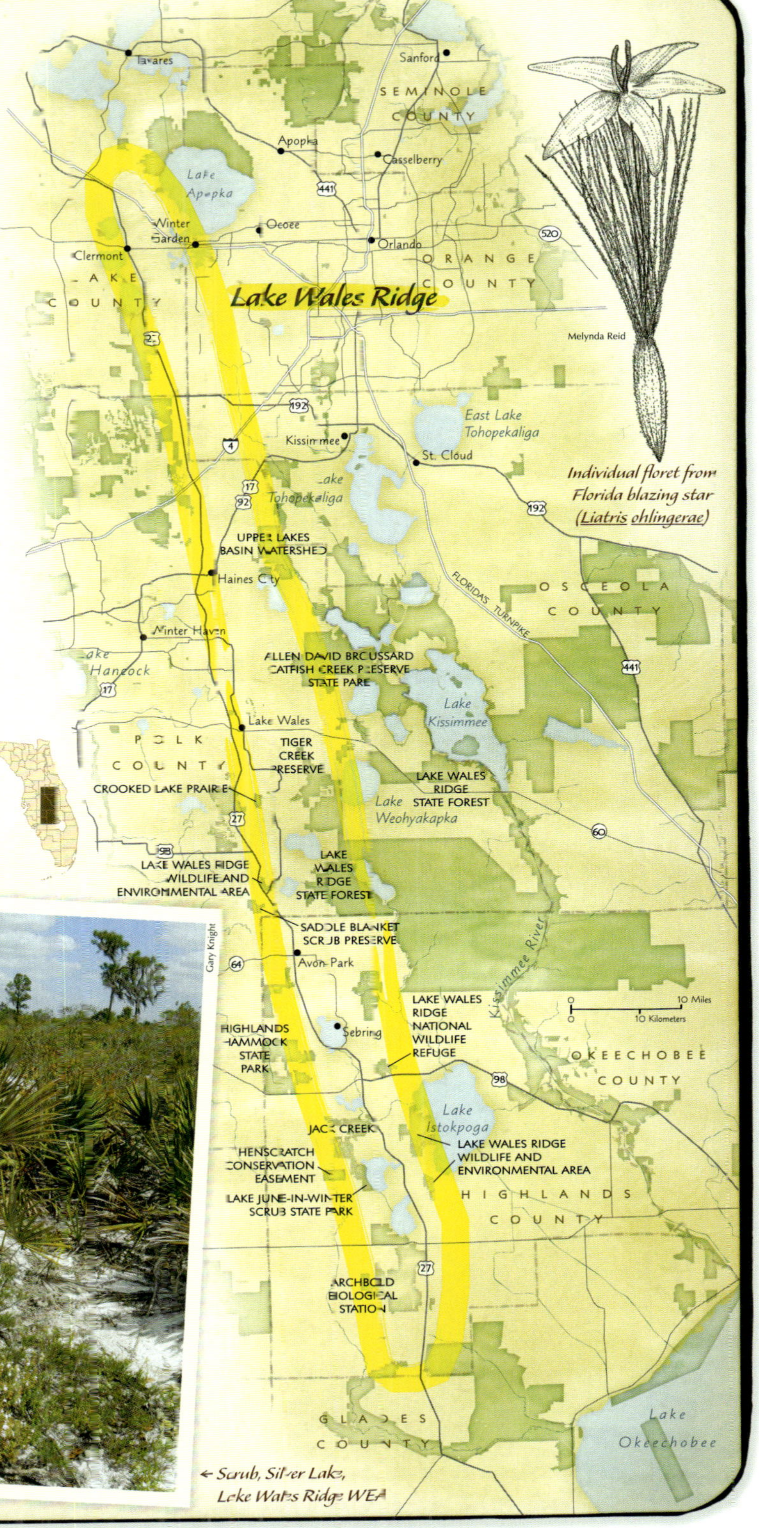

Individual floret from Florida blazing star (Liatris ohlingerae)

KEY ELEMENTS OF BIODIVERSITY

Natural Communities
Scrub

Plants

Cladonia perforata	perforate reindeer lichen
Crotalaria avonensis	Avon Park rabbit-bells
Dicerandra christmanii	Garrett's scrub balm
Dicerandra frutescens	scrub mint
Eryngium cuneifolium	wedge-leaved button-snakeroot
Euphorbia rosescens	scrub spurge
Lupinus aridorum	scrub lupine
Schizachyrium niveum	scrub bluestem
Warea amplexifolia	clasping warea
Ziziphus celata	scrub ziziphus

Animals

Aphelocoma coerulescens	Florida scrub-jay
Cicindela highlandensis	Highlands tiger beetle
Plestiodon egregius lividus	blue-tailed mole skink
Plestiodon reynoldsi	sand skink

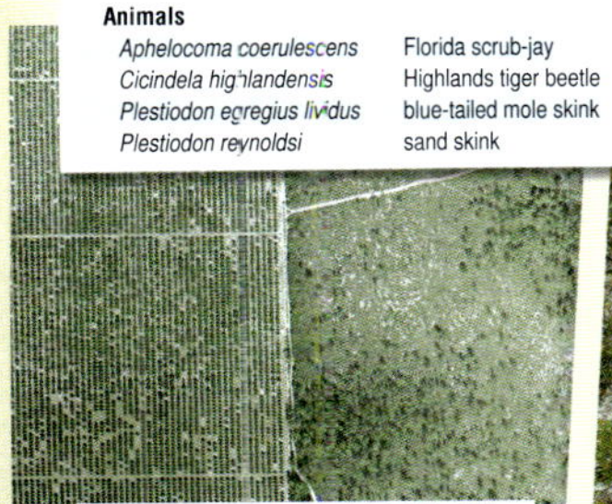

The sandy, well-drained soils of the Lake Wales Ridge proved highly suitable for citrus cultivation. Rows of citrus trees are clearly seen in this aerial photo.

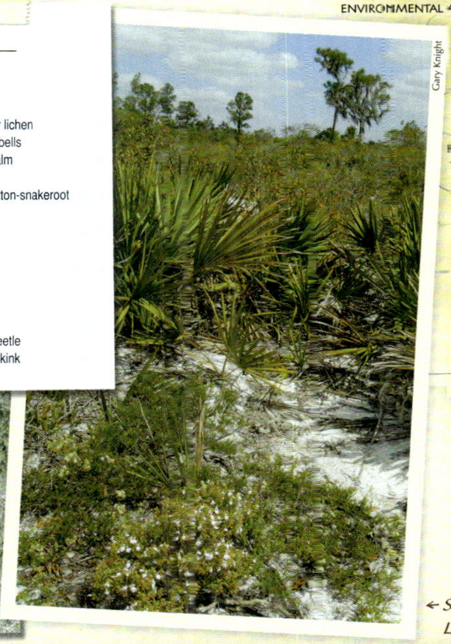

← *Scrub, Silver Lake, Lake Wales Ridge WEA*

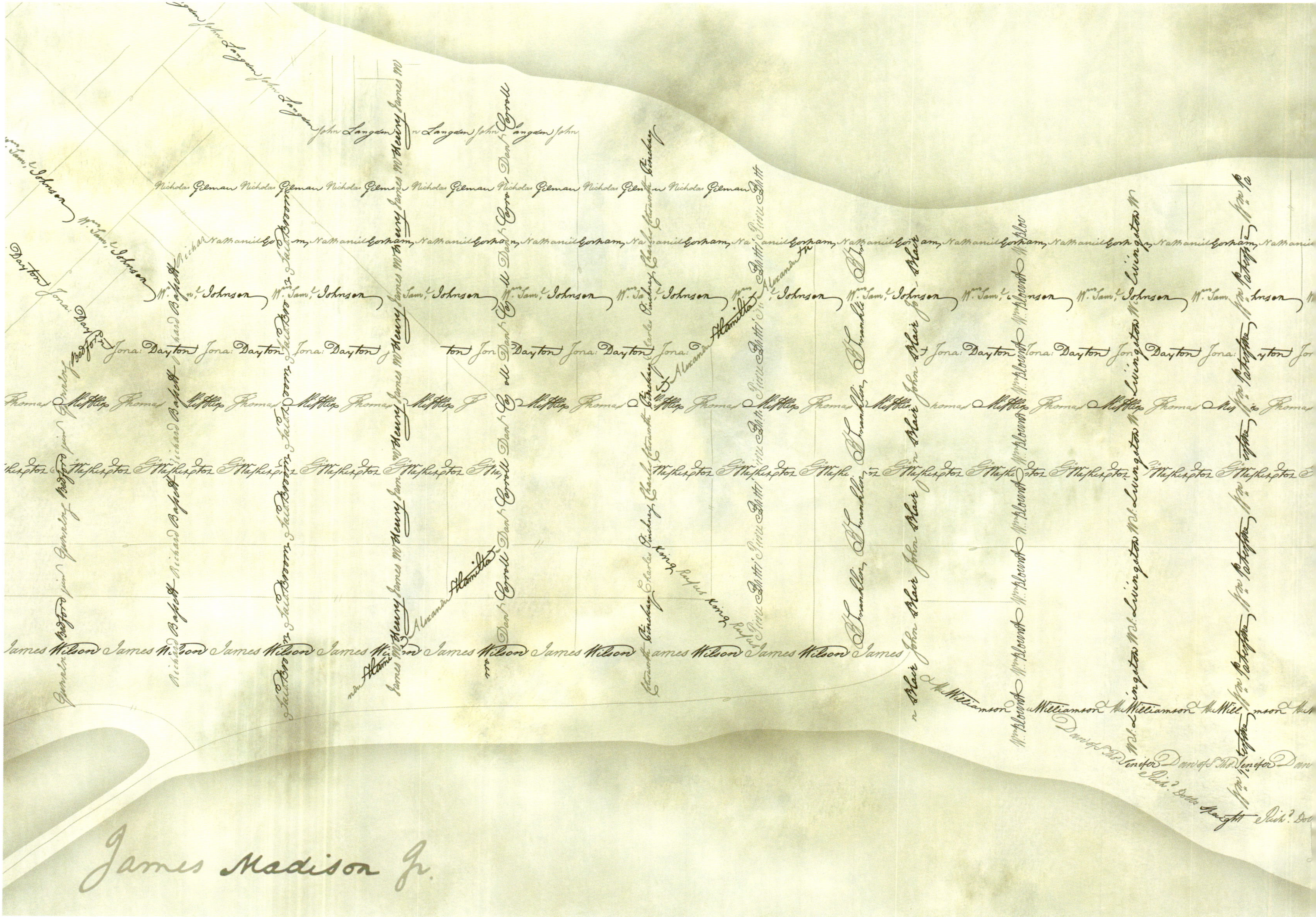
James Madison Jr.
John Langdon
Nicholas Gilman
Nathaniel Gorham
Wm. Saml. Johnson
Jona: Dayton
Thomas Mifflin
James Wilson
Richard Bassett
Gunning Bedford jun
Jacob Broom
James McHenry
Alexander Hamilton
Danl. Carroll
Charles Pinckney
Rufus King
Hu Williamson
Wm. Blount
Wm. Livingston
Wm. Paterson
Dan of St Thos. Jenifer
Richd. Dobbs Spaight
John Blair
B Franklin
Geo: Washington

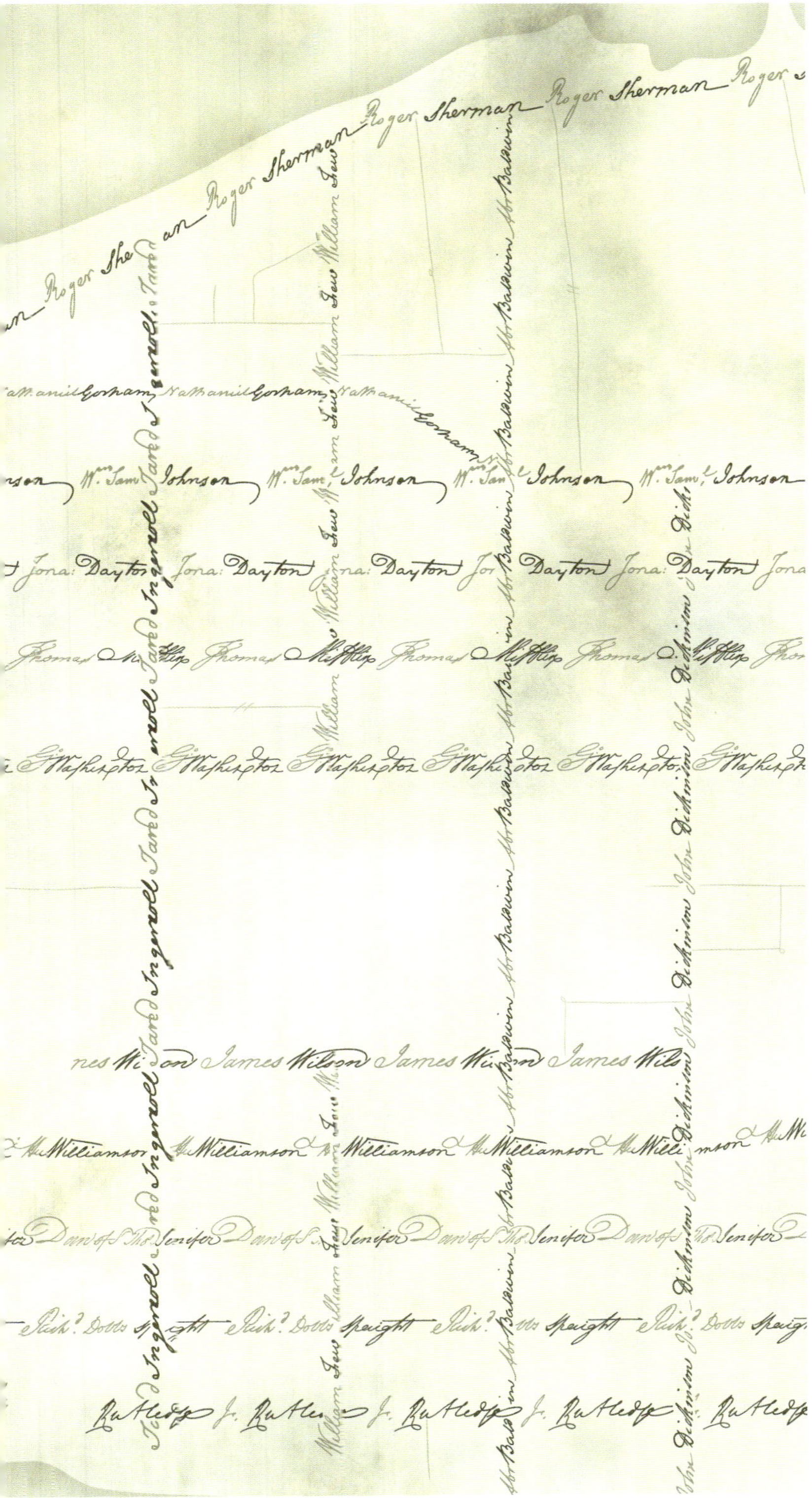

« The Ways of the Framers

Daniel P. Huffman — somethingaboutmaps

Everywhere we look, we see names on the landscape, meant to memorialize some historic figure deemed worthy of remembrance. But as generations pass, we rarely apprehend the significance of the fact that we live on *Adams* Street, or walk past *DeWaters* Hall on a trip through a campus. That's just what these landmarks are called; it doesn't even occur to us that they share names with specific human beings.

When James Doty laid out the first streets of Madison, Wisconsin, in 1836, he named them after the signers of the United States Constitution. Today, though, that connection is lost on many of the city's residents. I wanted to create something that would put readers in touch with the history embedded in their local geography.

Each street is rendered using the signature of the man for whom it is named, traced from a scan of the Constitution. Handwriting is deeply personal, and putting it on the map gives the reader a more direct, human connection with the historical figure attached to it. We do not merely see Washington's name; we see a little bit of his personality right there, tracing out the course of the broad avenue named for him. The signatures serve as a connection to persons long gone, and a reminder to readers that this past is embedded in the places they know and live.

The use of handwriting gives each street a distinct look, with its own individual character. We don't often see this sort of variation on maps, where it is standard to use uniform type styles for each carefully organized class of object. But I think that it becomes easier to forget the meaning of place names when their similarities are emphasized, rather than their differences. This map is meant to remind people that each has its own reason for being there, its own story.

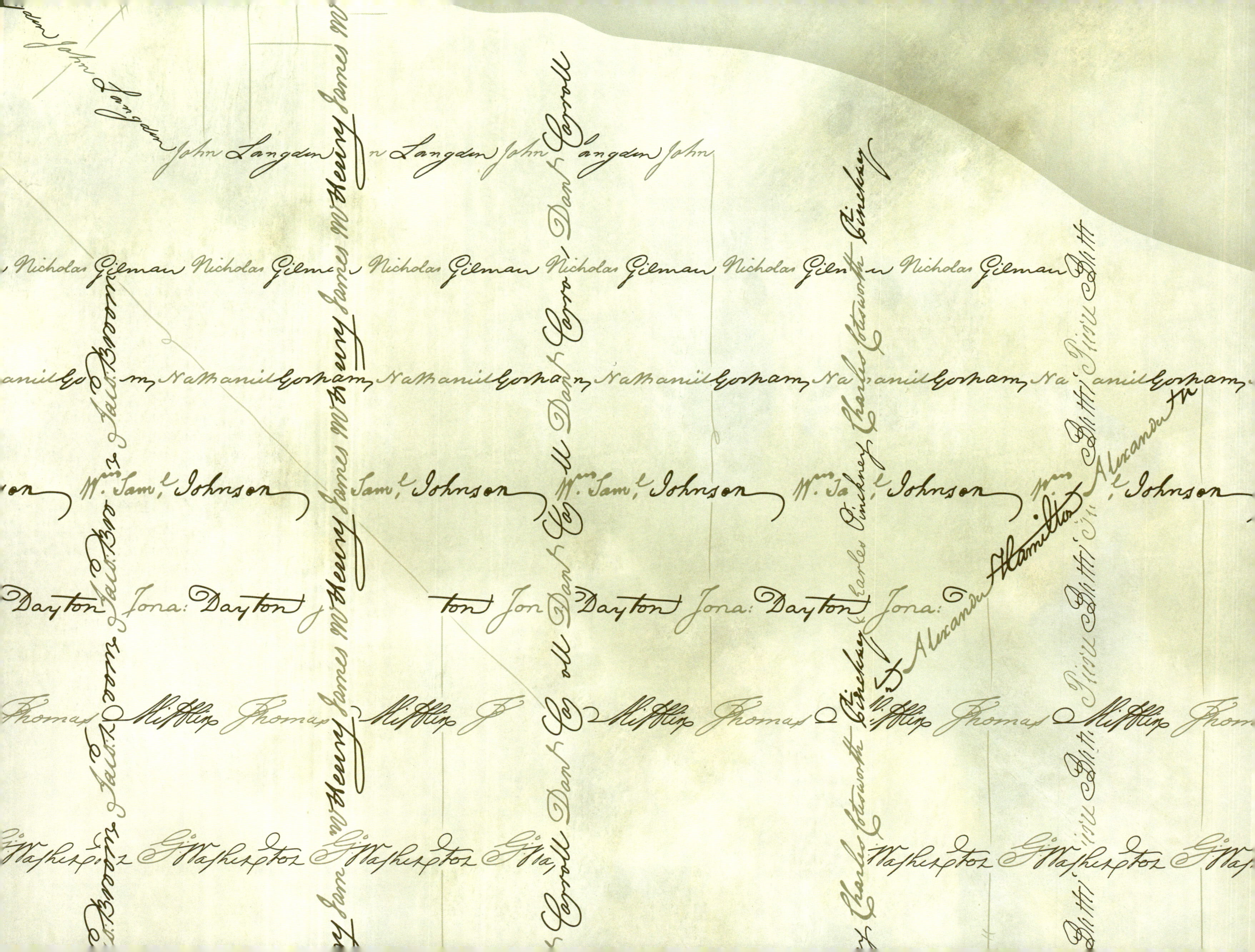

John Langdon
Nicholas Gilman
Nathaniel Gorham
Wm. Saml. Johnson
Jona: Dayton
Thomas Mifflin
Gᵒ. Washington
Jas. McHenry
Danl. Carroll
Charles Cotesworth Pinckney
Alexander Hamilton
Jacob Broom

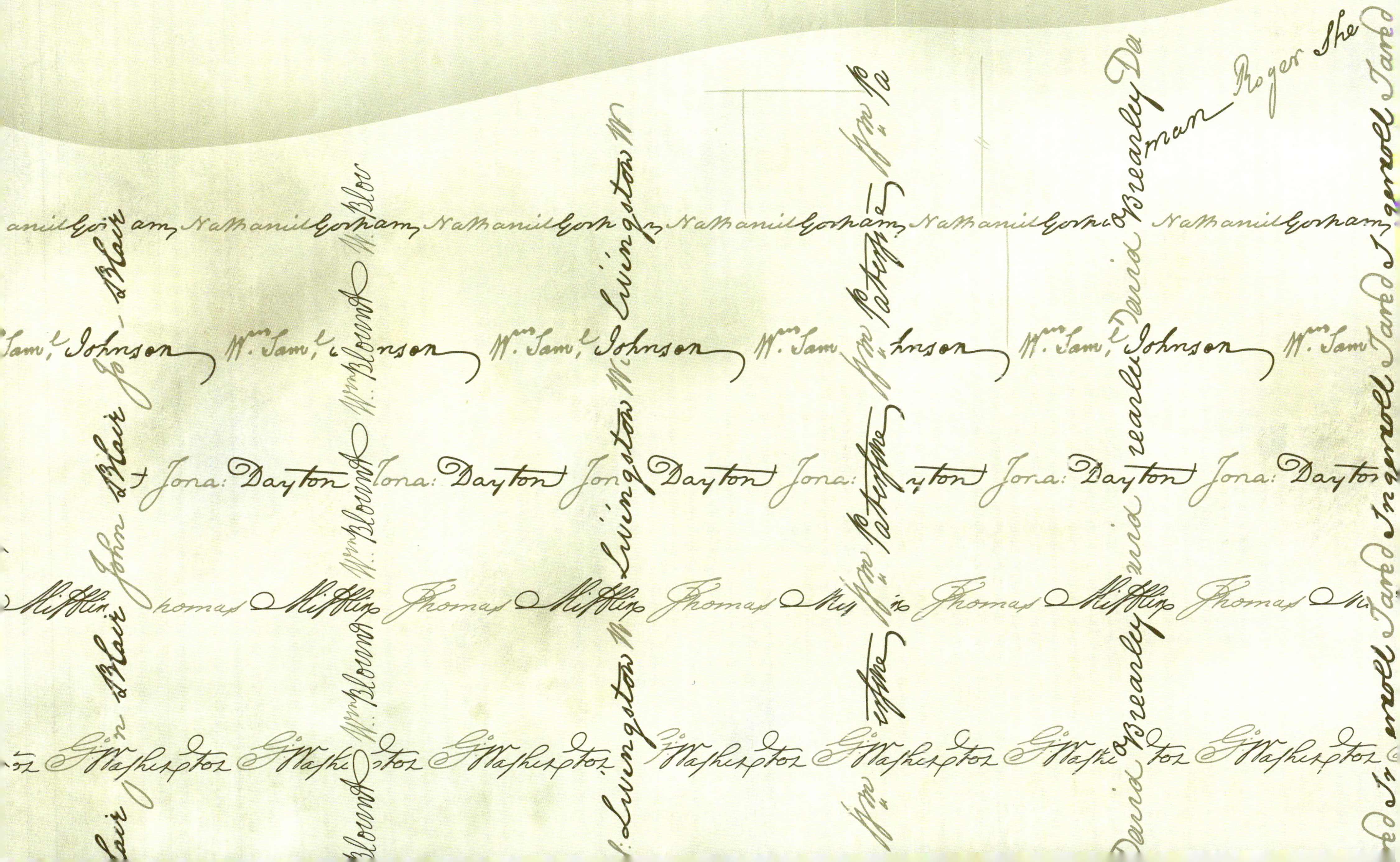

Nathaniel Gorham
Wm. Saml. Johnson
Jona: Dayton
Thomas Mifflin
Go: Washington
John Blair
Wm. Blount
Wm. Livingston
Wm. Paterson
David Brearley
Roger Sherman
Jared Ingersoll

D-Day Normandy »

David Deis — Dreamline Cartography

Produced for the June 2009 special D-Day issue of *America in WWII*, the purpose of this project was to assemble a fairly complete overview of the action on June 6, 1944. This double-page spread provided general support for numerous articles featured within the issue.

One of the biggest challenges associated with map design is figuring out exactly what needs to be shown. Maps in general—and military history maps in particular—often try to show too much information. As with all maps I work on, I tried to distill the contents in this map down to only what is absolutely necessary.

» *I tried to distill the contents in this map down to only what is absolutely necessary.*

I also needed this spread to roughly comply with the established look and feel of maps previously featured in *America in WWII* with respect to color, type, labeling conventions, etc. Over the years I have developed a map aesthetic for the magazine that features soft, muted raster backgrounds upon which more important thematic vector data and textual information are positioned. The idea is to make the map as clear and legible as possible.

This sort of design rationale may be traced back to my days as a cartography student, when I was taught to pay particular attention to the concepts of retinal variables and visual levels. That is, maps and other information graphics should endeavor to establish a hierarchy of information, with the more important elements demanding greater visual attention. Contrasting "soft" backgrounds with sharp, colorful, thematic data is one way I attempt to achieve that hierarchy.

Since the coverage area here includes a great deal of water, I felt it was critical to find the right ocean treatment that would be strong enough to establish good figure/ground, and subtle enough to handle a lot of text and thematic information. The lightly textured, pastel-colored ocean was the result. The coastal fade effect provided a good finishing touch, as it helped with the figure/ground relationship and lightened up the coastline areas where much thematic detail is concentrated.

My ultimate goal here was to produce a composition that presents a lot of information in a manner that does not make the composition feel too cluttered or complicated.

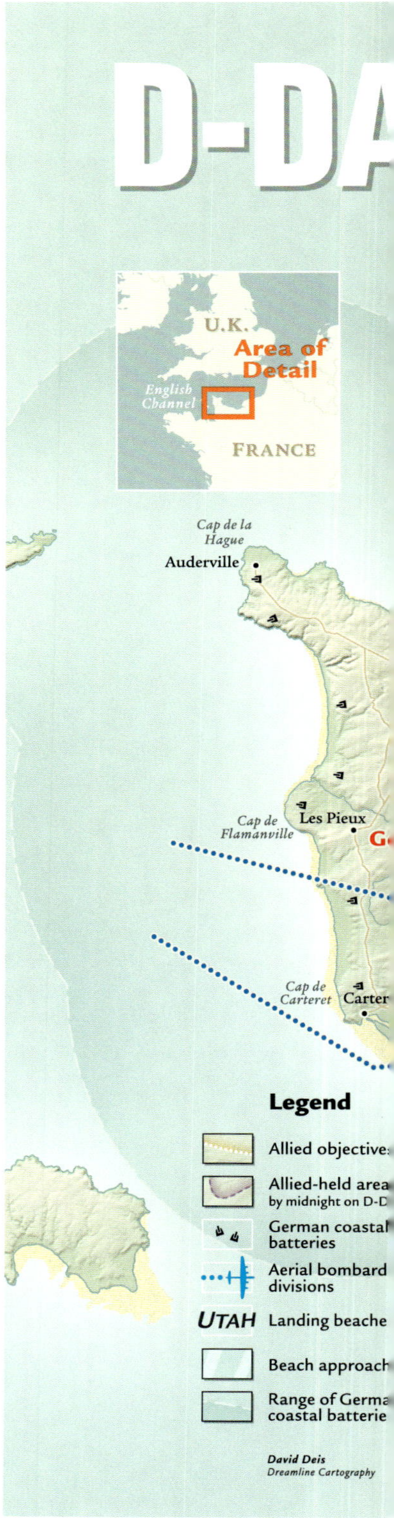

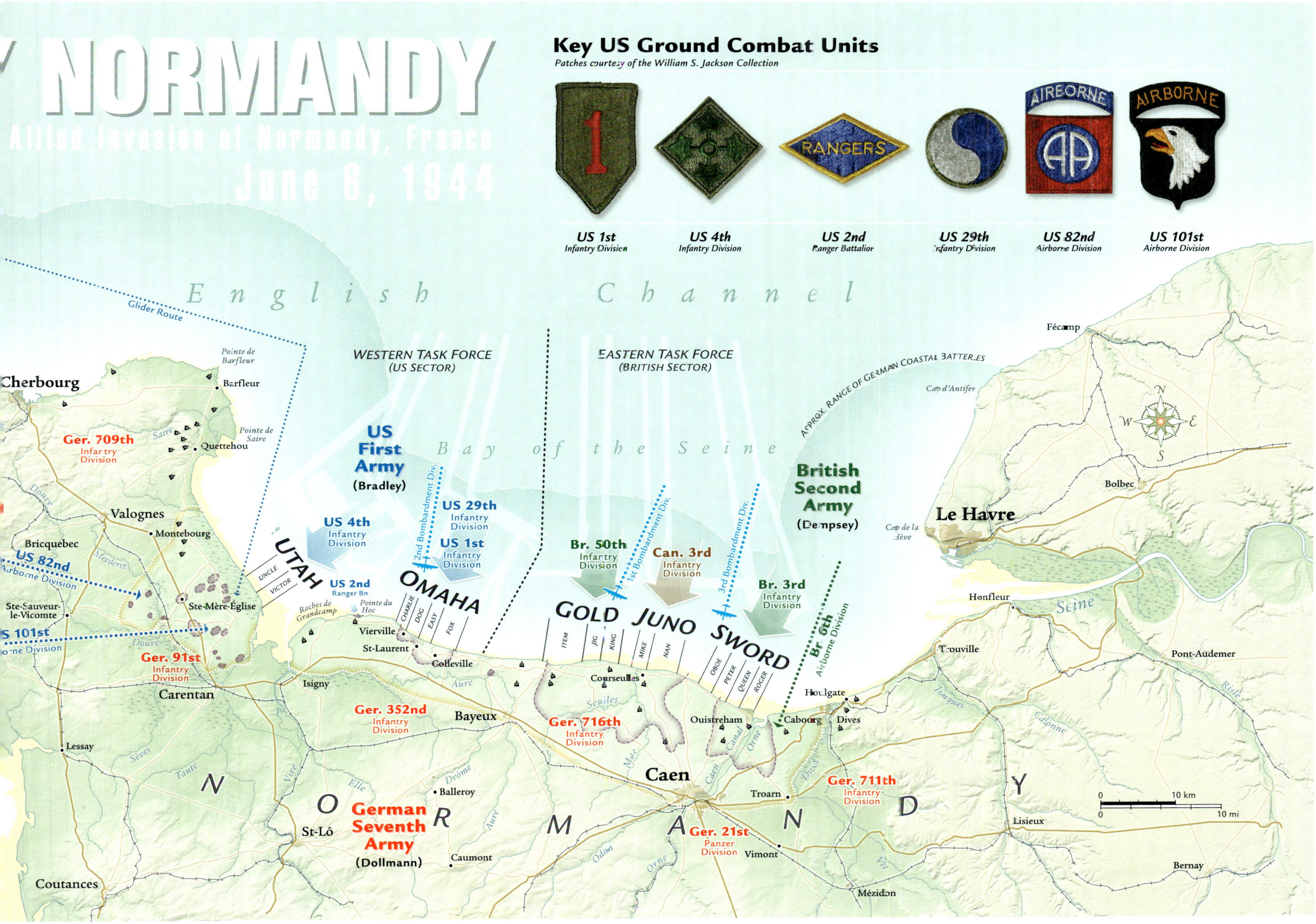
NORMANDY
Allied Invasion of Normandy, France
June 6, 1944
Key US Ground Combat Units
Patches courtesy of the William S. Jackson Collection
US 1st Infantry Division
US 4th Infantry Division
US 2nd Ranger Battalion
US 29th Infantry Division
US 82nd Airborne Division
US 101st Airborne Division
English Channel
Bay of the Seine
Glider Route
WESTERN TASK FORCE (US SECTOR)
EASTERN TASK FORCE (BRITISH SECTOR)
APPROX. RANGE OF GERMAN COASTAL BATTERIES
US First Army (Bradley)
British Second Army (Dempsey)
US 4th Infantry Division
US 29th Infantry Division
US 1st Infantry Division
US 2nd Ranger Bn
2nd Bombardment Div.
1st Bombardment Div.
3rd Bombardment Div.
Br. 50th Infantry Division
Can. 3rd Infantry Division
Br. 3rd Infantry Division
Br. 6th Airborne Division
US 82nd Airborne Division
US 101st Airborne Division
UTAH
UNCLE
VICTOR
OMAHA
CHARLIE
DOG
EASY
FOX
GOLD
ITEM
JIG
KING
JUNO
MIKE
NAN
SWORD
OBOE
PETER
QUEEN
ROGER
Ger. 709th Infantry Division
Ger. 91st Infantry Division
Ger. 352nd Infantry Division
Ger. 716th Infantry Division
Ger. 711th Infantry Division
Ger. 21st Panzer Division
German Seventh Army (Dollmann)
Cherbourg
Pointe de Barfleur
Barfleur
Pointe de Saire
Quettehou
Saire
Valognes
Montebourg
Bricquebec
Merderet
Douve
Ste-Sauveur-le-Vicomte
Ste-Mère-Église
Roches de Grandcamp
Pointe du Hoc
Vierville
St-Laurent
Colleville
Isigny
Aure
Carentan
Lessay
Sèves
Taute
Vire
Elle
Drôme
Balleroy
St-Lô
Caumont
Coutances
Bayeux
Courseulles
Seulles
Mue
Ouistreham
Canal
Orne
Caen
Odon
Troarn
Vimont
Mézidon
Cabourg
Houlgate
Dives
Honfleur
Seine
Trouville
Touques
Calonne
Pont-Audemer
Risle
Lisieux
Bernay
Le Havre
Cap de la Hève
Cap d'Antifer
Fécamp
Bolbec
N
S
W
E
N O R M A N D Y
0 10 km
0 10 mi

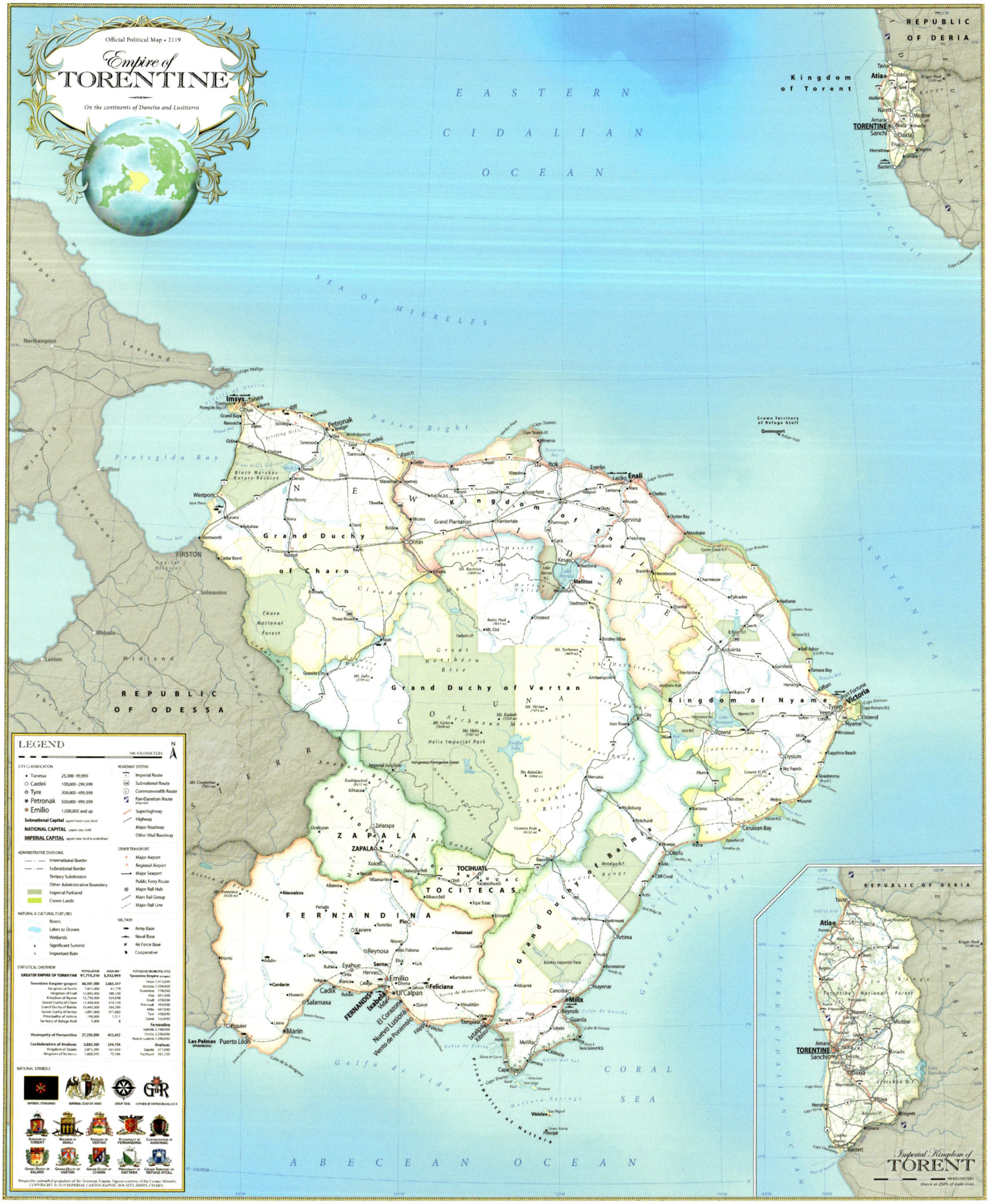

Official Political Map • 2119
Empire of
TORENTINE
On the continents of Danetia and Lusitierra
EASTERN CIDALIAN OCEAN
SEA OF MIERELES
ESSLYAN SEA
CORAL SEA
ABECEAN OCEAN
Pasco Bight
Protegida Bay
Golfo de Vida
Grand Coral Bight
Kingdom of Torent
REPUBLIC OF DERIA
REPUBLIC OF ODESSA
Crown Territory of Refuge Atoll
Queensport
Kingdom of Enali
Grand Duchy of Charn
Grand Duchy of Vertan
Kingdom of Nyame
Grand Duchy of Balmix
FERNANDINA
ZAPALA
TOCITECAS
Principality of Hattara
Imsys
Petronak
Enali
Victoria
Emilio
Milix
TORENTINE
Atia
Sanchi
Las Palmas
Puerto León
Cape Town
Imperial Kingdom of
TORENT
LEGEND

« Empire of Torentine: A Political Map

Brian E. Stoll

Being a rather impressionable teen, I didn't end up following my dreams of becoming a cartographer or an urban planner. So I guess you could say this map of the Torentine Empire is my first "grown-up" map, and a way to fill that void. It is actually the third step in a transitional project. I say transitional because I grew up spending untold hours drawing maps by hand, and, while I do have experience in digital art, I never had attempted a cartographic project this extensive on the computer. So Torentine was my muse for satisfying that transition which, thanks to decades of raster and vector career work, only took about three passes to refine.

My goal with the map was to convey in a realistic, professional, and attractive way, this entirely made-up empire. I wanted to give a sense of its unique internal political geography and position in the world while including some vital statistical and cultural data. In short, I wanted viewers to feel as if they were looking at a place that might actually exist.

Originally I envisioned something less modern, probably because I was attracted to the printing effects of older atlases. But I opted for the crisp, clean colors you see here to keep the map legible and to showcase the colorful heraldry. Legibility was probably the greatest challenge, as I wanted to hint at complexity without overburdening the viewing experience.

By referencing post-WWII maps and atlases, I learned that balance and stylistic consistency were key to a successfully informative, yet uncluttered map. After considerable trial and error I was able to achieve this to my satisfaction through strategic text placement, a harmonious color palette, and rational size and placement of various elements.

Two elements I feel contribute to the appeal of the map are the cartouche and the heraldry. The 3D globe cartouche is decorative, but also serves to give a relative scale and location of the nation in the wider world. The heraldry, also a bit of an indulgence, was intended to add depth to individual realms of the empire.

In short, I wanted viewers to feel as if they were looking at a place that might actually exist.

LEGEND

500 KILOMETERS

N

CITY CLASSIFICATION

- Tianessa 25,000–99,999
- Cardeii 100,000–299,999
- Tyre 300,000–499,999
- Petronak 500,000–999,999
- Emilio 1,000,000 and up

Subnational Capital upper/lower case, bold

NATIONAL CAPITAL upper case, bold

IMPERIAL CAPITAL upper case, bold & underlined

ROADWAY SYSTEM

- Imperial Route
- Subnational Route
- Commonwealth Route
- Pan-Danetian Route proposed
- Superhighway
- Highway
- Major Roadway
- Other Vital Roadway

ADMINISTRATIVE DIVISIONS

- International Border
- Subnational Border
- Tertiary Subdivision
- Other Administrative Boundary
- Imperial Parkland
- Crown Lands

OTHER TRANSPORT

- Major Airport
- Regional Airport
- Major Seaport
- Public Ferry Route
- Major Rail Hub
- Main Rail Group
- Major Rail Line

NATURAL & CULTURAL FEATURES

- Rivers
- Lakes or Oceans
- Wetlands
- Significant Summit
- Important Ruin

MILITARY

- Army Base
- Naval Base
- Air Force Base
- Cooperative

STATISTICAL OVERVIEW

	POPULATION	AREA KM²
GREATER EMPIRE OF TORENTINE	**97,715,210**	**3,332,955**
Torentine Empire (proper)	**66,581,000**	**2,665,347**
Kingdom of Torent	7,614,000	41,779
Kingdom of Enali	15,043,000	386,530
Kingdom of Nyame	12,736,000	424,838
Grand Duchy of Charn	17,449,000	474,729
Grand Duchy of Balmix	10,443,000	364,269
Grand Duchy of Vertan	3,097,000	971,883
Principality of Hattara	198,000	1,311
Territory of Refuge Atoll	1,000	8
Viceroyalty of Fernandina	**27,250,850**	**433,452**
Confederation of Anahuac	**3,883,360**	**234,156**
Kingdom of Zapala	2,875,290	161,050
Kingdom of Tocitecas	1,008,070	73,106

POPULOUS MUNICIPALITIES	
Torentine Empire (proper)	
Imsys	1,412,000
Victoria	1,109,000
Torentine	778,000
Atia	651,000
Enali	478,000
Petronak	463,000
Milix	447,000
Tyre	438,000
Daxia	322,000
Fernandina	
Ixabela	2,748,000
Emilio	2,556,000
Nuevo Lusiora	1,396,000
Anahuac	
Zapala	377,000
Tocihuatl	301,250

NATIONAL SYMBOLS

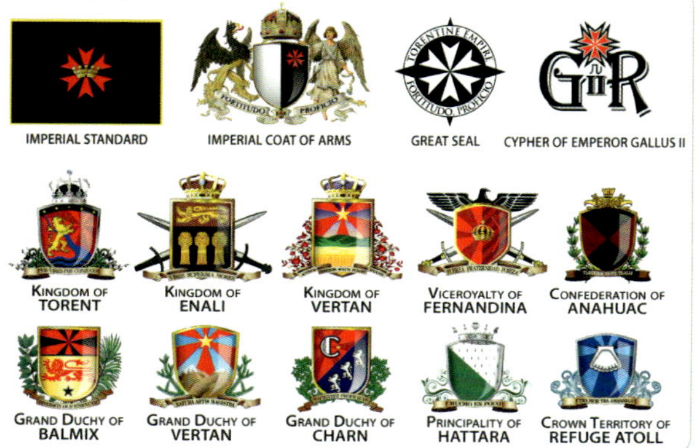

Perspective azimuthal projection of the Torentine Empire. Figures courtesy of the Census Ministry.

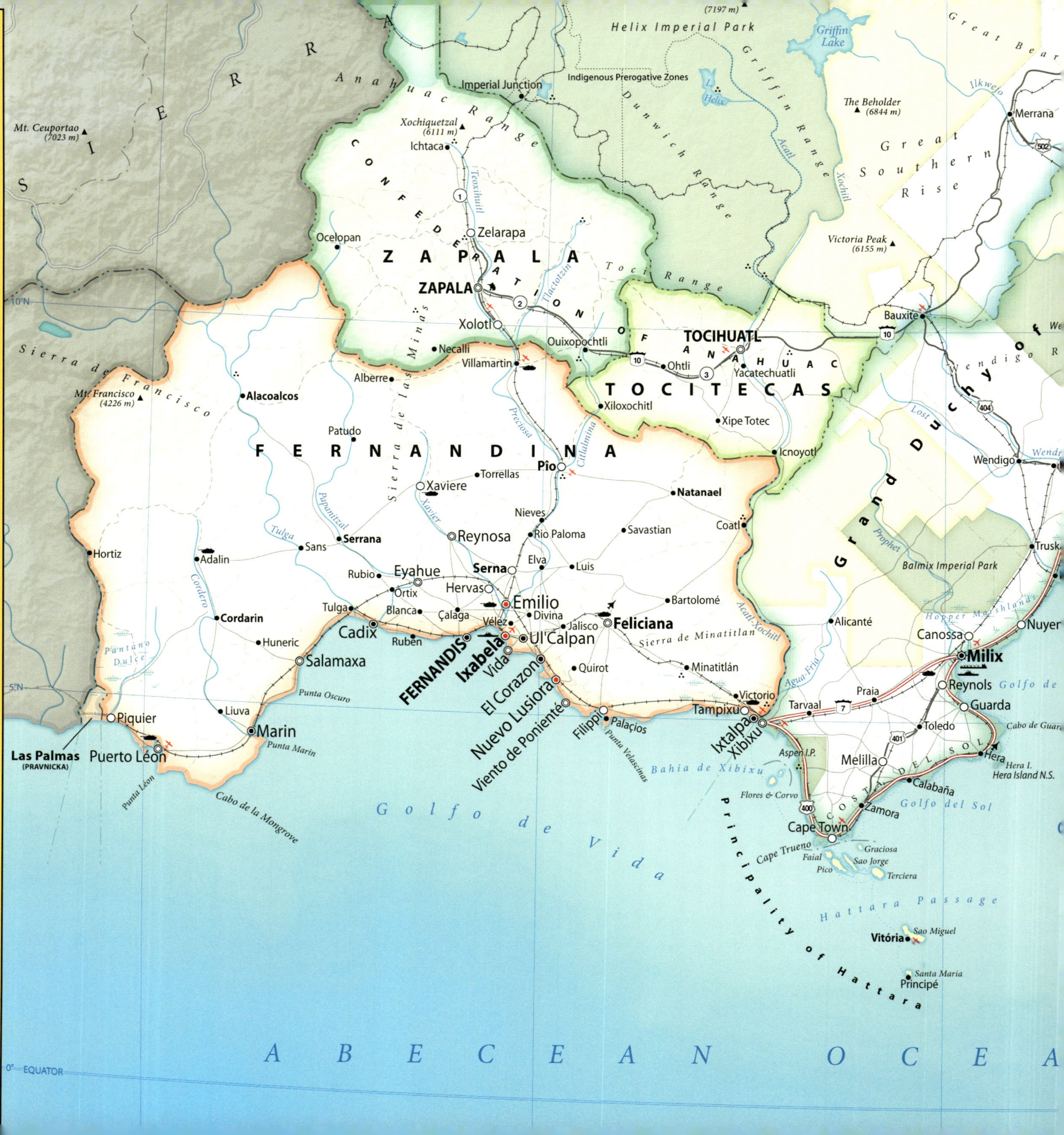

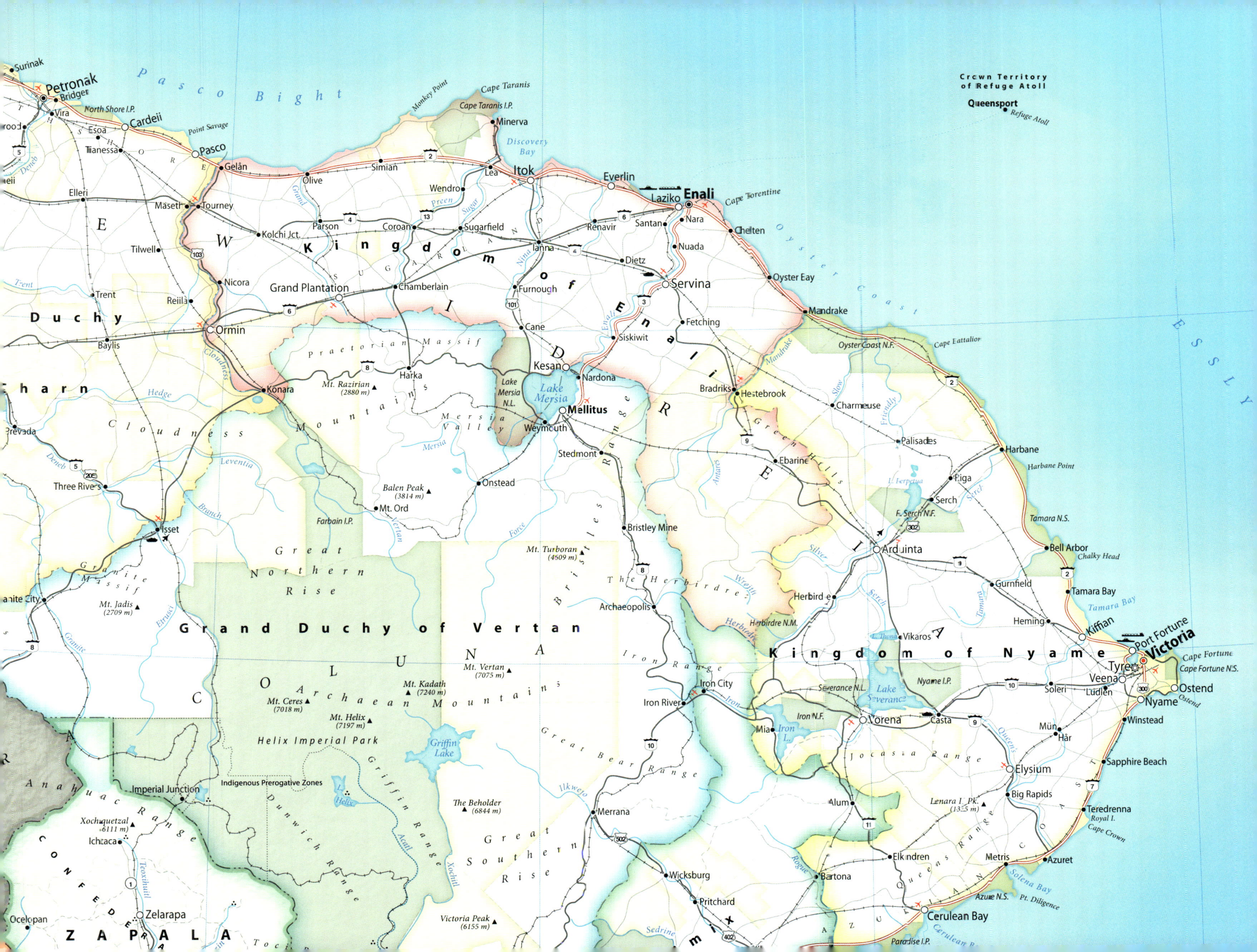

Crown Territory of Refuge Atoll
Queensport
Refuge Atoll
Pasco Bight
Surinak
Petronak
Bridger
Vira
North Shore I.P.
Cardeii
Esoa
Tianessa
Point Savage
Pasco
Gelân
Monkey Point
Cape Taranis
Cape Taranis I.P.
Minerva
Discovery Bay
Simian
Olive
Lea
Itok
Wendro
Everlin
Laziko
Enali
Cape Torentine
Nara
Chelten
Nuada
Santan
Renavir
Dietz
Oyster Bay
Oyster Coast
Servina
Elleri
Maseti
Tourney
Kolchi Jct.
Parson
Coroan
Sugarfield
Ianna
Tilwell
Kingdom of Enali
Sugar Land
Nicora
Grand Plantation
Chamberlain
Furnough
Trent
Reilla
Duchy
Ormin
Baylis
Cane
Fetching
Siskiwit
Mandrake
Oyster Coast N.F.
Cape Eattalior
Essly
Praetorian Massif
Harka
Kesan
Nardona
Konara
Mt. Razirian (2880 m)
Mountains
Lake Mersia
Lake Mersia N.L.
Mellitus
Weymouth
Mersia Valley
Mersia
Bradriks
Hestebrook
Charmeuse
Palisades
Harbane
Harbane Point
Cloudness
Hedge
Leventia
Prevada
Deneb
Three Rivers
Stedmont
Ebarine
Green Hills
Piga
Serch
L. Perpetua
F. Serch N.F.
Tamara N.S.
Onstead
Balen Peak (3814 m)
Mt. Ord
Farbain I.P.
Branch
Verlan
Force
Isset
Bristley Mine
Ardquinta
Bell Arbor
Chalky Head
Great Northern Rise
Granite Massif
Mt. Turboran (4609 m)
Bristles Range
The Herbirdre
Gurnfield
Tamara Bay
Archaeopolis
Herbird
Herbirdre N.M.
Heming
Kiffian
Port Fortune
Victoria
Cape Fortune
Cape Fortune N.S.
Granite City
Mt. Jadis (2709 m)
Grand Duchy of Vertan
Iron Range
Vikaros
Kingdom of Nyame
Tyre
Veena
Ludlen
Soleri
Ostend
Nyame
Winstead
Mt. Vertan (7075 m)
Mt. Kadath (7240 m)
Mt. Ceres (7018 m)
Mt. Helix (7197 m)
Archaean Mountains
Helix Imperial Park
Griffin Lake
Iron City
Iron River
Severance N.L.
Lake Severance
Nyame I.P.
Iron N.F.
Vorena
Casta
Mia
Iron L.
Mün
Hàr
Sapphire Beach
Jocasta Range
Elysium
Big Rapids
Great Bear Range
Imperial Junction
Indigenous Prerogative Zones
Anahuac Range
Xochiquetzal (6111 m)
Ichtaca
Dunwich Range
Griffin Range
The Beholder (6844 m)
Merrana
Alum
Lenara I. Pk. (1325 m)
Teredrenna
Royal I.
Cape Crown
Great Southern Rise
Elkindren
Metris
Azuret
Queens Range
Wicksburg
Bartona
Solena Bay
Azure N.S.
Pt. Diligence
Pritchard
Cerulean Bay
Paradise I.P.
Confederacion
Zelarapa
Ocelopan
Zapala
Victoria Peak (6155 m)
Sedrine

Japan's Swirling Seas »

Virginia Mason — National Geographic Magazine

» *Many maps focus on terrestrial features, so this was an opportunity to work on a design that features water.*

This map accompanied a *National Geographic Magazine* feature story on the biodiversity of marine life in Japan's surrounding ocean and seas (November, 2010). The brilliant colors of the photographs by Brian Skerry presented an opportunity for a map that compliments, but does not overwhelm, the story color palette.

National Geographic Magazine publishes many different kinds of maps, from small reference to story-rich, thematic maps. While most are designed to stand alone, they must also fit the look, feel, and design of the story in which they reside. In other words, the editorial cartography of *National Geographic Magazine* must hold its own as an element of story making, while complementing the text and photographic narrative of feature stories.

Scientists at the Geophysical Fluid Dynamics Laboratory at NOAA created a high-resolution model of sea surface temperatures for this project. I reviewed several scenes in order to find one that showed warm and cold eddies in a compositionally agreeable place, and that allowed for explanatory text. I selected a simulation depicting a time of year that highlights the contrast in temperatures most clearly. I also wanted to pick a dataset that gave the best impression of the currents' motion.

Many maps focus on terrestrial features, so this was an opportunity to work on a design that features water and pushes terrestrial features into negative space. Reversing to black what might traditionally be white negative space makes the bright colors pop to the foreground.

This map reminds me that there are so many things on this planet that we can't see with our eyes, but they are there all the same. Sometimes, science makes the visualization of nature possible.

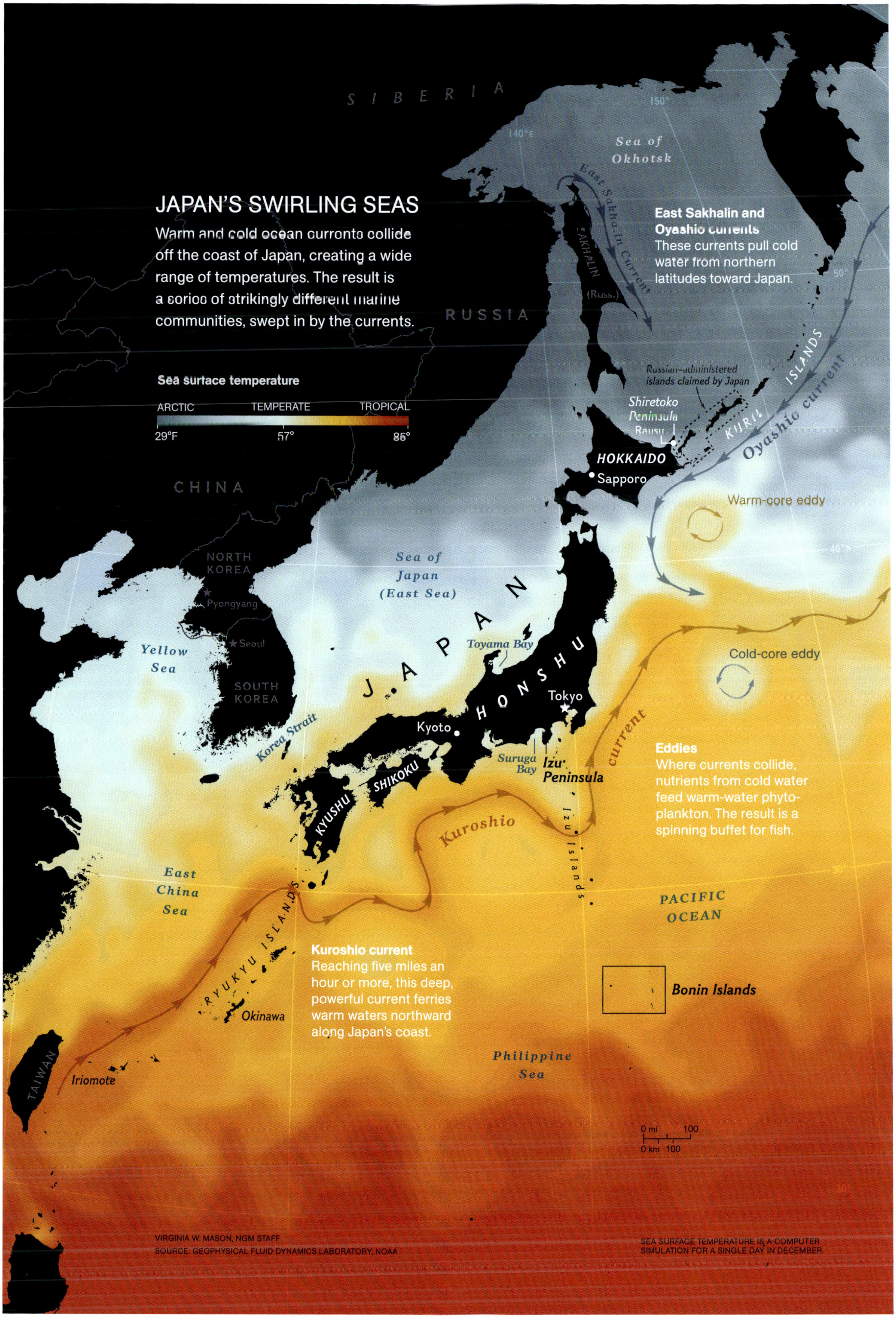

JAPAN'S SWIRLING SEAS
Warm and cold ocean currents collide off the coast of Japan, creating a wide range of temperatures. The result is a series of strikingly different marine communities, swept in by the currents.
Sea surface temperature
ARCTIC
TEMPERATE
TROPICAL
29°F
57°
85°
SIBERIA
RUSSIA
CHINA
NORTH KOREA
Pyongyang
Seoul
SOUTH KOREA
Yellow Sea
Sea of Okhotsk
SAKHALIN
(Russ.)
East Sakhalin Current
East Sakhalin and Oyashio currents
These currents pull cold water from northern latitudes toward Japan.
Russian-administered islands claimed by Japan
Shiretoko Peninsula
Rausu
HOKKAIDO
Sapporo
KURIL ISLANDS
Oyashio current
Warm-core eddy
Sea of Japan (East Sea)
JAPAN
Toyama Bay
HONSHU
Tokyo
Kyoto
Cold-core eddy
Korea Strait
Suruga Bay
Izu Peninsula
current
Eddies
Where currents collide, nutrients from cold water feed warm-water phyto-plankton. The result is a spinning buffet for fish.
SHIKOKU
KYUSHU
Kuroshio
Izu Islands
East China Sea
PACIFIC OCEAN
RYUKYU ISLANDS
Kuroshio current
Reaching five miles an hour or more, this deep, powerful current ferries warm waters northward along Japan's coast.
Bonin Islands
Okinawa
Iriomote
TAIWAN
Philippine Sea
0 mi 100
0 km 100
140°E
150°
50°
40°N
30°
VIRGINIA W. MASON, NGM STAFF
SOURCE: GEOPHYSICAL FLUID DYNAMICS LABORATORY, NOAA
SEA SURFACE TEMPERATURE IS A COMPUTER SIMULATION FOR A SINGLE DAY IN DECEMBER.

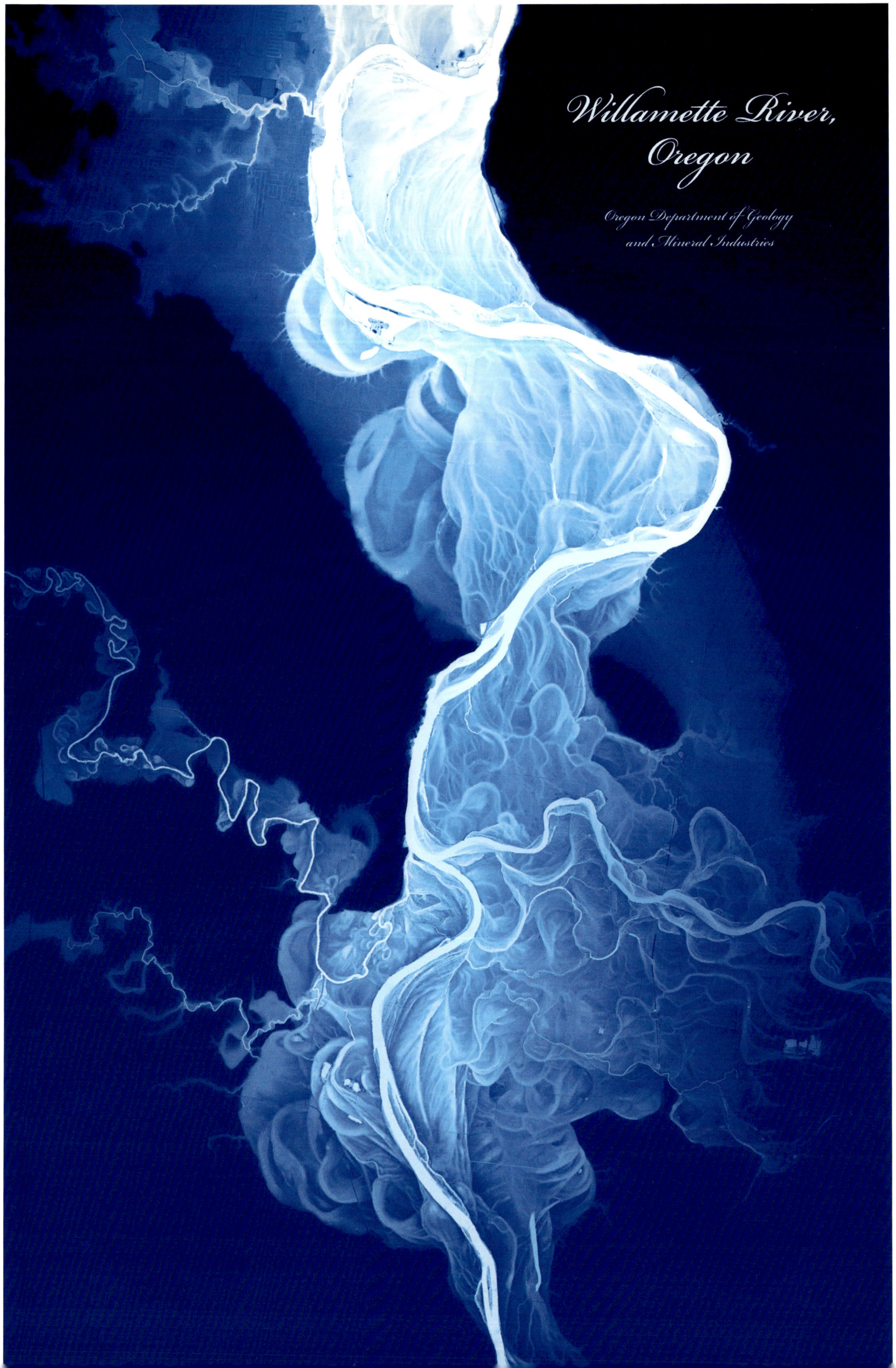
Willamette River,
Oregon
Oregon Department of Geology
and Mineral Industries

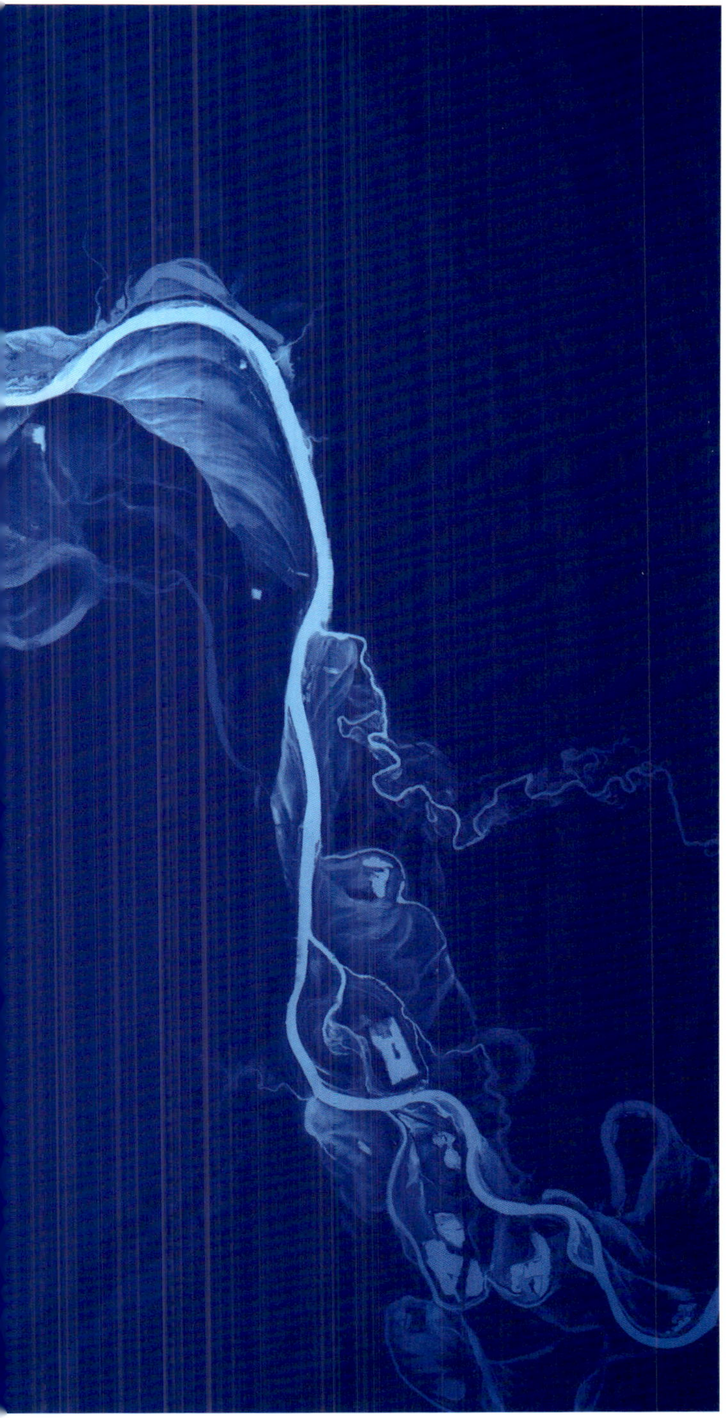

« Willamette River

Daniel E. Coe — Oregon Department of Geology and Mineral Industries

This image seems to be half map, half painting; somewhere between science and art. It materialized one day while I was working on new ways of visualizing lidar data. I had recently discovered and found inspiration in Mississippi River maps from the 1944 Army Corps of Engineers study, the *Geological Investigation of the Alluvial Valley of the Lower Mississippi River*, by Harold N. Fisk. These maps, I thought, stood alone as beautifully abstract works of art, yet they contained an immense body of scientific detail at the same time. Perhaps with these map images floating somewhere in the back of my mind, I searched for new cartographic ways of interpreting the streams of the Willamette Valley.

I manipulated a digital elevation model of the Willamette River to display about a 50-foot elevation range, from low elevations (displayed in white) fading to higher elevations (displayed in dark blue). From the scientific perspective, this effect visually replaced the relatively flat landscape of the valley floor with vivid historical channels, showing the dynamic movements the river has made in recent millennia. From the artistic point of view, the map/painting took on a completely different quality, ranging from electrical storm to nervous system.

For a short while, I tried adding cartographic context by including roads, cities, or counties. Every time I added new elements, however, it felt like something was being taken away from the image by obscuring it. While the techniques I used to create this image have since led to newer physiographic maps of the Willamette Valley, I decided that this version of the Willamette River should stand alone in its natural artistic beauty.

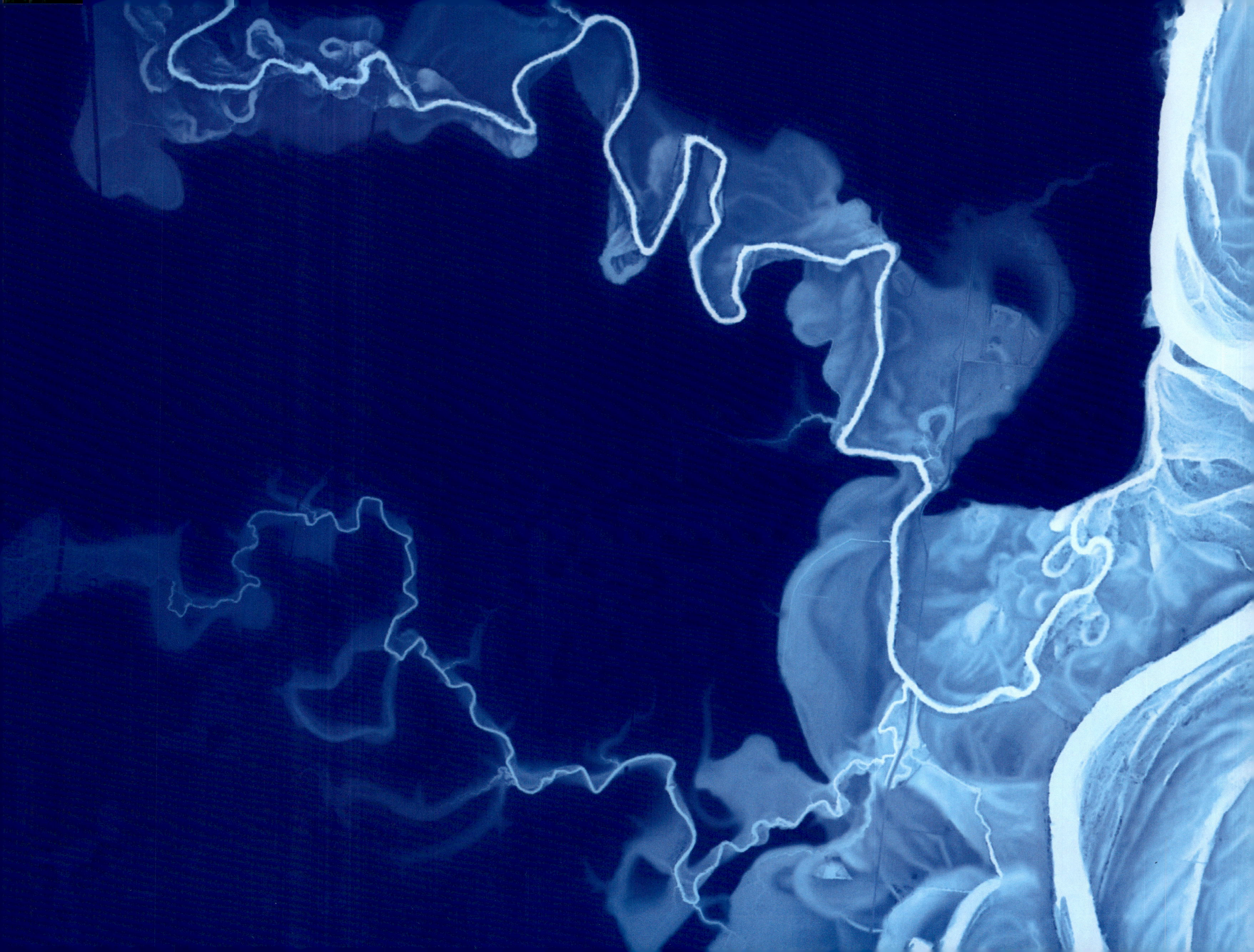

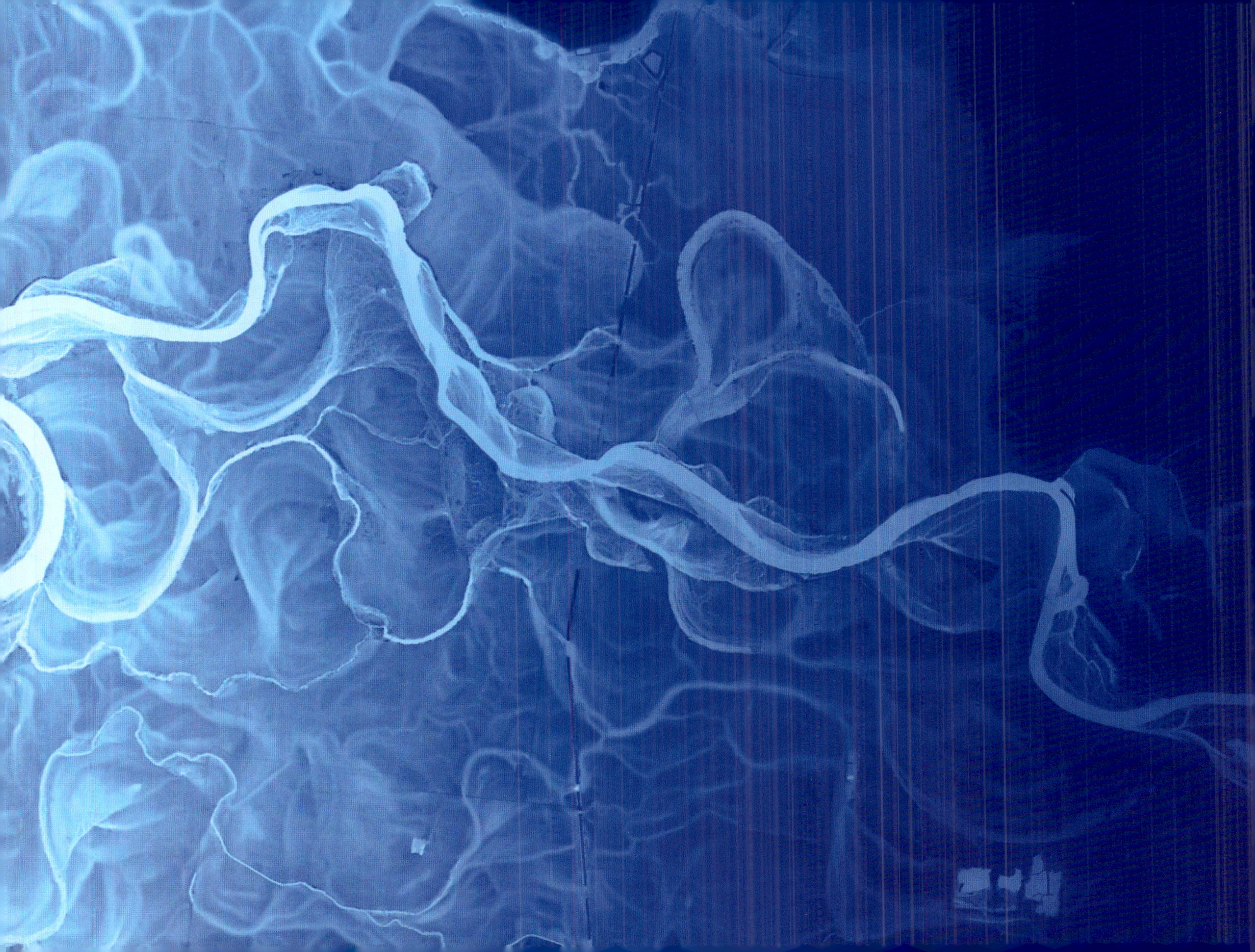

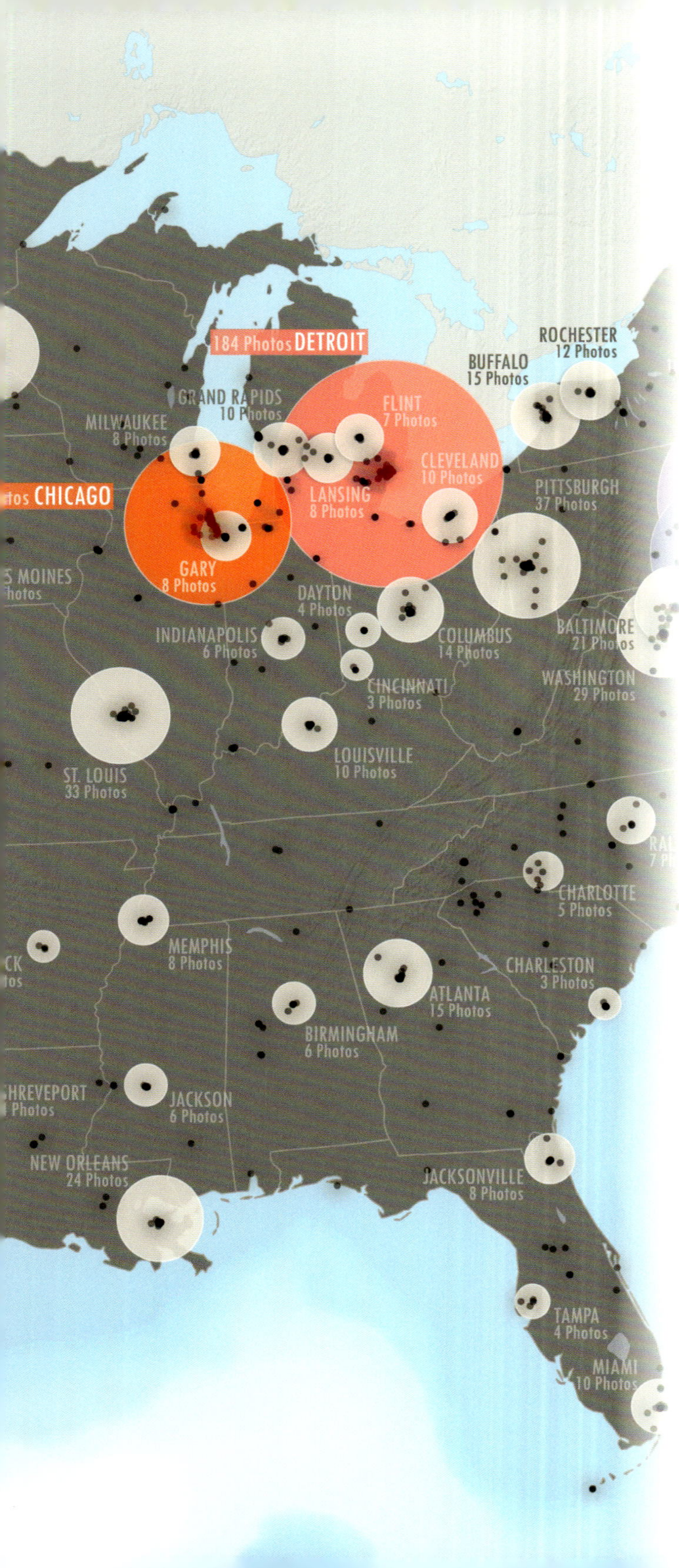

Picturing Urban Decay »

Derek Watkins

This map examines perceptions of urban decay in the United States through images uploaded by users of the photo-sharing site Flickr. "Urban ruins" have become a popular photographic subject in recent years, and I was interested in uncovering the geography of people's perceptions of ugliness on the landscape. This map attempts to explore such themes.

I aimed for a clean and restrained aesthetic that allowed the data to shine through. Subdued refinement can quickly become stark minimalism, however, so I also included some details to give the map more visual substance, such as subtle drop shadows, shaded relief, and bathymetry layers. While technically unnecessary in terms of data communication, I felt these elements add to the overall effect of the final product.

I chose a dark brown color for the individual data points to convey the feeling of blight spreading across the landscape. Similar reasoning led me to set the title and body type in Day Roman. This font echoes the imperfections of early printing techniques, and provides a look of flawed refinement that works well with my subject matter. The drab background color that I chose also helps to communicate a sense of decline.

I provided visual contrast to this depressing backdrop in the form of bright, almost-cheerful colors for the most photographed urban areas. This added visual interest to the map while also highlighting some of the most interesting stories that emerged from the data. I carried those same colors into the accompanying data graphics, visually binding together stories about individual cities across the page.

Picturing Urban Decay

... an exploration of blight as depicted through online photography.

"Urban decay" refers to deteriorating remains of abandoned or derelict urban areas. Usually associated with large cities in the United States and Europe, the concept is a hard one to define. Scholars generally point to segregation, deindustrialization, and population loss as the factors behind decay, but popular portrayals in the United States often approach the subject in aesthetic terms. Each person sees the world slightly differently than their neighbor, and something that one finds unattractive may go unnoticed by the next. The concept of urban decay is arguably as much a social perception as it is a consequence of economic decline, but perceptions are difficult to pin down precisely.

In recent years, millions of people have begun to collectively publish unprecedented amounts of information online. Through a variety of media such as blogs, commentaries, videos, and photographs, connected societies perpetually contribute to a growing accumulation of data. Exploring this mass can allow insights into underlying perceptions that otherwise may have remained isolated in the minds of the authors. If users attach information linking their contributions to specific locations, trends can be mapped to reveal broad-scale patterns. Users have uploaded over 3 million such "geotagged" images to the photo-sharing website flickr.com.

Mapping the distribution of photos of urban decay that have been uploaded to flickr reveals the places which people find to be the most compelling examples of blight in the United States.

The result is a map that illustrates the distribution of decay, as defined by the culturally-informed aesthetic preferences of hundreds of individual photographers.

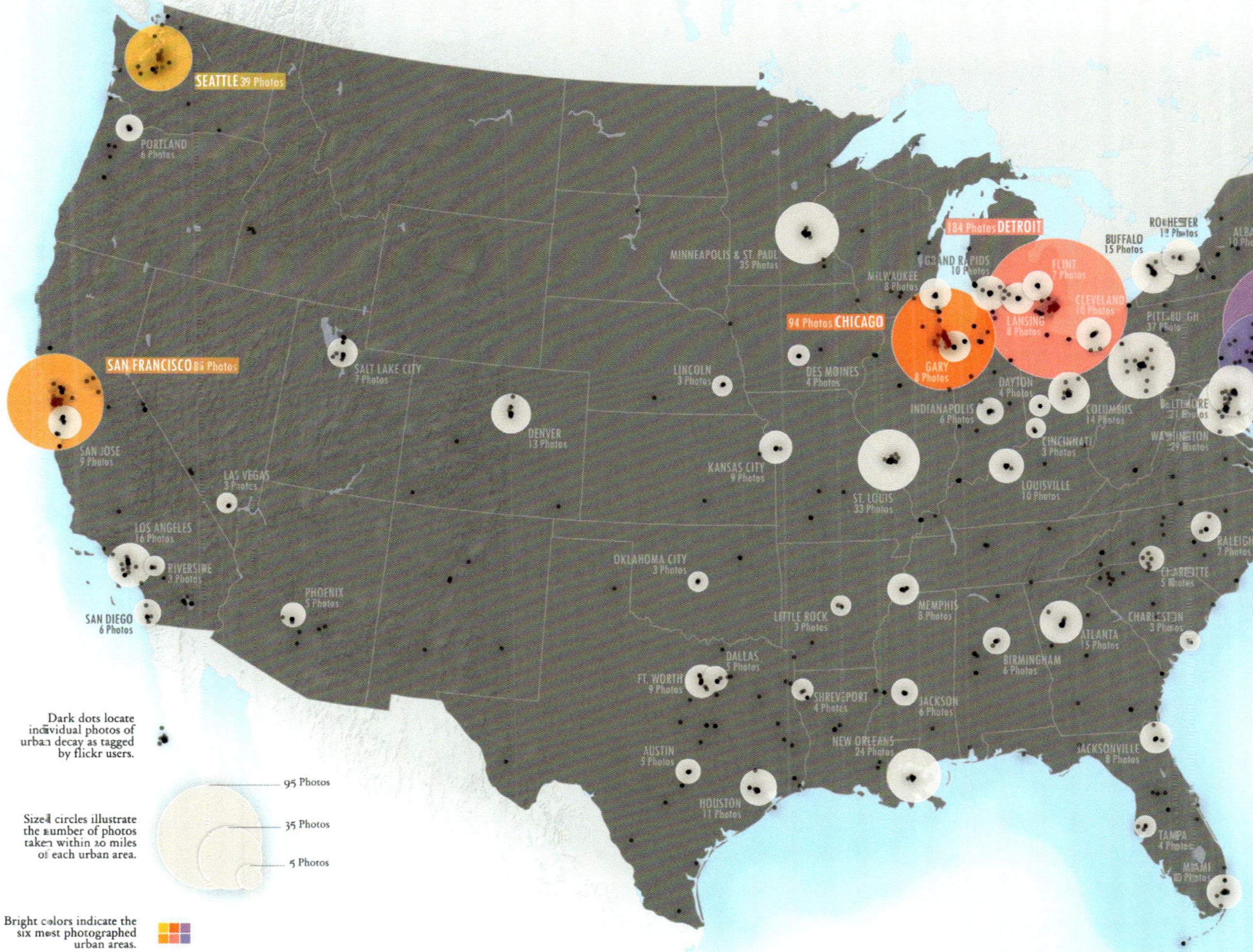

Causes and Consequences of Decay

Urban decay is a complex phenomenon, and the cyclical feedback between its causes and effects can be hard to decipher. Population change, employment in manufacturing, and home values are three measures that can predict or reflect the emergence of decay.

Population Change

Urban areas change over time to accommodate the ebb and flow of population. Urban decay can emerge as a consequence of population loss through the desertion of residences and workplaces. Conversely, blight can often spur people to leave urban areas in greater numbers. People generally move according to their economic ability to do so, which often leaves disadvantaged groups behind to deal with the consequences of decline.

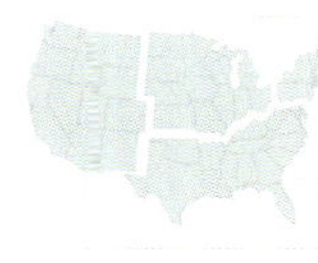

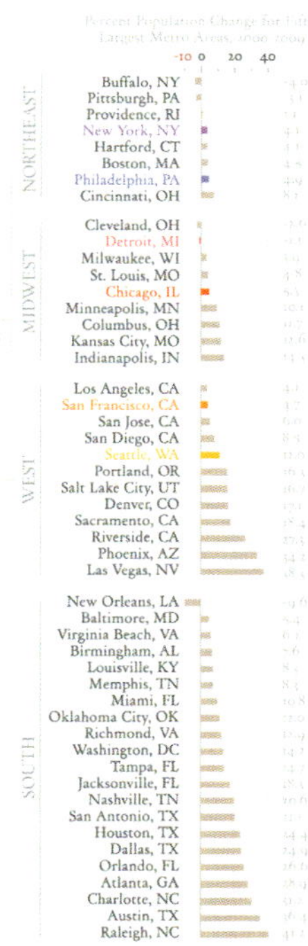

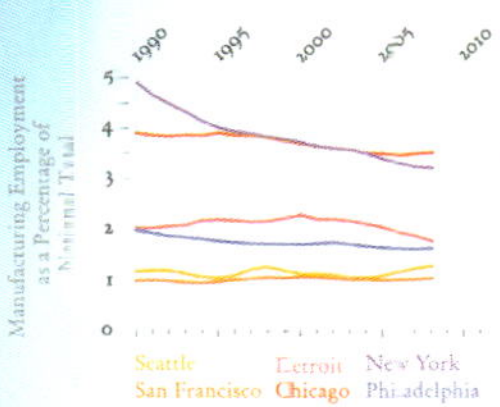

Manufacturing Employment

Popular perceptions attribute decay to deindustrialization. Derelict factories often become "modern ruins" that are popular photographic subjects. Historical changes in the distribution of the manufacturing workforce can give some insight into patterns of real and percieved decay.

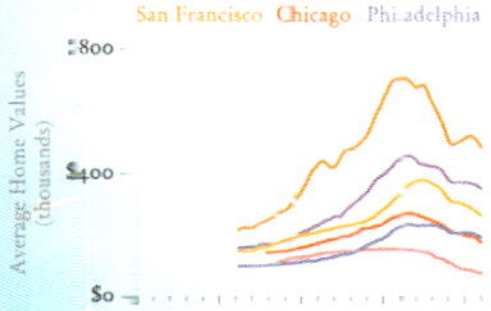

Home Values

Among many other variables, home values reflect public perceptions of the "quality" of a place. Declining, low-value residential areas can themselves become photographic subjects of decay. The stigma of noteriety attached to blighted places can become a self-sustaining, driving home values lower and quickening decline.

Derek Watkins, 2011

Data from Flickr, Zillow, Natural Earth, the U.S. Census Bureau, and the U.S. Bureau of Labor Statistics.

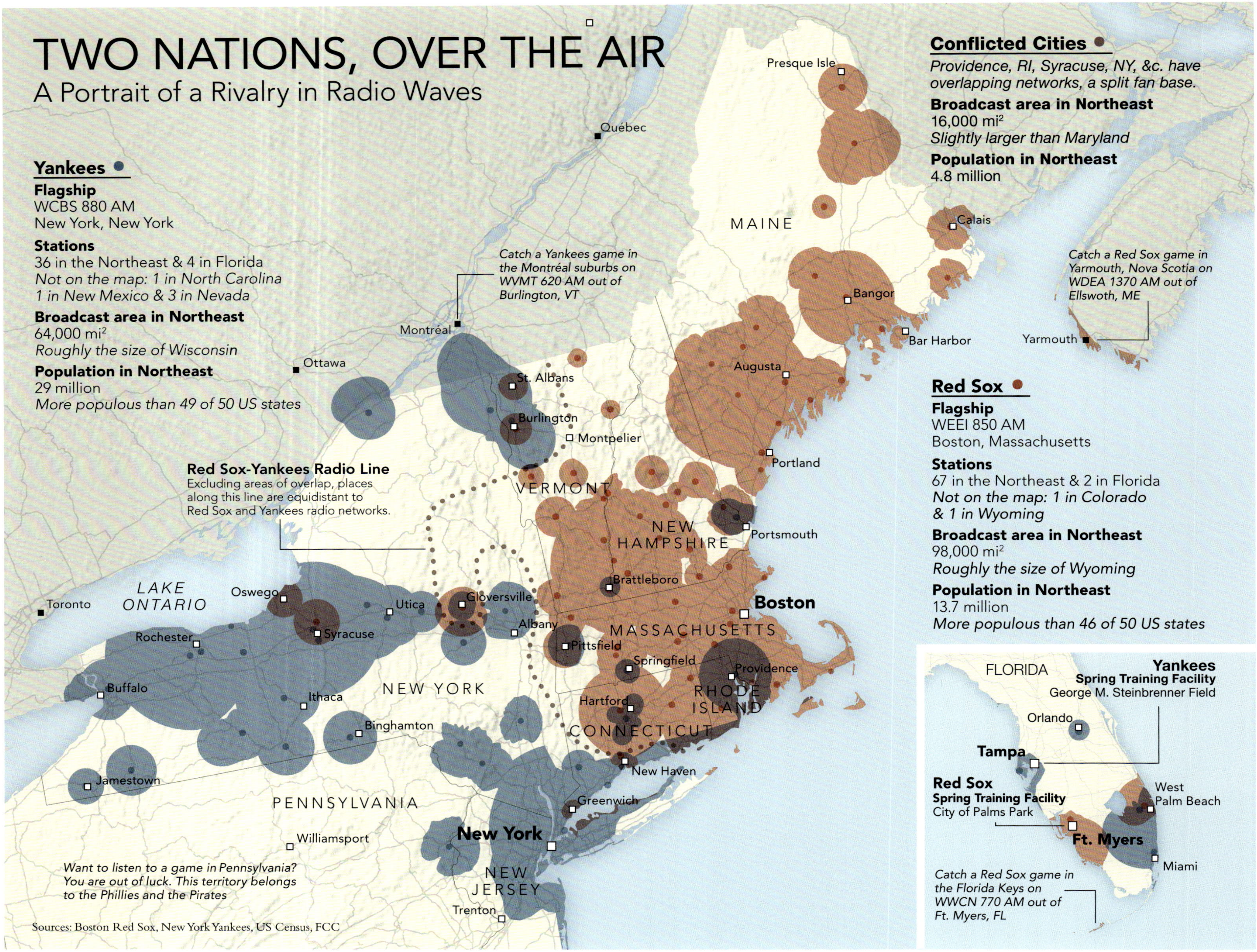
TWO NATIONS, OVER THE AIR
A Portrait of a Rivalry in Radio Waves
Yankees
Flagship
WCBS 880 AM
New York, New York
Stations
36 in the Northeast & 4 in Florida
Not on the map: 1 in North Carolina
1 in New Mexico & 3 in Nevada
Broadcast area in Northeast
64,000 mi²
Roughly the size of Wisconsin
Population in Northeast
29 million
More populous than 49 of 50 US states
Conflicted Cities
Providence, RI, Syracuse, NY, &c. have overlapping networks, a split fan base.
Broadcast area in Northeast
16,000 mi²
Slightly larger than Maryland
Population in Northeast
4.8 million
Red Sox
Flagship
WEEI 850 AM
Boston, Massachusetts
Stations
67 in the Northeast & 2 in Florida
Not on the map: 1 in Colorado
& 1 in Wyoming
Broadcast area in Northeast
98,000 mi²
Roughly the size of Wyoming
Population in Northeast
13.7 million
More populous than 46 of 50 US states
Catch a Yankees game in the Montréal suburbs on WVMT 620 AM out of Burlington, VT
Catch a Red Sox game in Yarmouth, Nova Scotia on WDEA 1370 AM out of Ellswoth, ME
Red Sox-Yankees Radio Line
Excluding areas of overlap, places along this line are equidistant to Red Sox and Yankees radio networks.
Want to listen to a game in Pennsylvania? You are out of luck. This territory belongs to the Phillies and the Pirates
Québec
Presque Isle
MAINE
Calais
Bangor
Bar Harbor
Yarmouth
Augusta
Portland
Portsmouth
Montréal
Ottawa
St. Albans
Burlington
Montpelier
VERMONT
NEW HAMPSHIRE
Brattleboro
Boston
MASSACHUSETTS
Pittsfield
Springfield
Providence
RHODE ISLAND
Hartford
CONNECTICUT
New Haven
Greenwich
Toronto
LAKE ONTARIO
Oswego
Rochester
Syracuse
Utica
Gloversville
Albany
Buffalo
Ithaca
NEW YORK
Binghamton
Jamestown
PENNSYLVANIA
Williamsport
New York
NEW JERSEY
Trenton
FLORIDA
Yankees
Spring Training Facility
George M. Steinbrenner Field
Orlando
Tampa
Red Sox
Spring Training Facility
City of Palms Park
West Palm Beach
Ft. Myers
Miami
Catch a Red Sox game in the Florida Keys on WWCN 770 AM out of Ft. Myers, FL
Sources: Boston Red Sox, New York Yankees, US Census, FCC

Still, the cultural landscapes of New England that have been draped with the radio waves of Sox and Yankees games for decades remain.

« Two Nations, Over the Air

Timothy R. Wallace — University of Wisconsin–Madison

In New England, it's a summer tradition to hit the road with the car radio tuned to a Red Sox game. As the broadcast slowly crackles out of range, fans scan the airwaves to find another station carrying the game, and another—until suddenly, the Sox are off the dial and the only game on the air is something else entirely: the Yankees playing ball.

From an early age, I used car trips like this as an opportunity to learn the landscape, and this map was the inevitable byproduct. I stared at accordion-folded road maps for ages while carefully tweaking the radio dial in barely-perceptible increments. Over the years, my mind developed a rich cognitive map of where baseball lives in the airwaves over the region.

Nowadays, we can listen to any game we want anywhere we want. Still, the cultural landscapes of New England that have been draped with the radio waves of Sox and Yankees games for decades remain. This map aims to visualize these landscapes, beginning with a subtle terrain that shows where radio waves—and humans—may be met with physical resistance. Broadcast areas are shown with a mostly transparent cloud of reception, rather than a hard boundary.

Radio waves are challenging to map because signal strengths vary based on numerous and sometimes-ephemeral factors. To compensate, this map shows conservative estimates of areas likely to have a constant signal. This was a purposeful design decision; attempting to map areas with reported occasional reception, such as Nova Scotia and Chicago, would have resulted in a more muddled look.

Mostly devoid of ornamentation, with sans serif type, it is intended to be a clean and legible radio reference map. The areas with clear reception of Red Sox and Yankees games are depicted in alternate colors, and areas where both teams can be heard are mixed accordingly.

Simply mapping radio coverage areas by team, however, would not satisfy true fans. So, with data extrapolated from broadcast areas, I mapped the Red Sox-Yankees line, a hypothetical boundary between radio landscapes. Cross the line and things may not look different—but they just might sound different.

» ... *we need to think more carefully about how we map social statistics, since the way we make maps influences the way we think about spatial planning and design.*

A Taxonomy of Transitions »

Bill Rankin — Yale University

Any city-dweller knows that most neighborhoods don't have clear boundaries. Yet on maps, neighborhoods are almost always drawn as perfectly bounded areas with sharp edges—homogeneous zones of ethnicity or class. This is especially true for maps of Chicago, a self-described "city of neighborhoods" notorious for its segregation.

This map takes a fresh look at the cartography of neighborhoods through a dot-density technique. The approach is relatively simple: each dot maps 25 people. Colors are based on racial self-identification from the United States Census. Dot-density maps have been around for more than a century, but they're still relatively rare. Typically, racial or ethnic data is shown using a choropleth map, with different areas shaded using solid colors. This kind of mapping, however, makes a strong visual argument that I want to avoid. Choropleth maps in newspapers or atlases usually only show which group holds a majority within a given area; this means that these maps implicitly reinforce political ideals of ethnic homogeneity and national self-determination. When solid colors are used to map city neighborhoods, they make a similar argument. Neighborhoods end up looking like miniature territorial states.

By using dots instead of solid colors, my map immediately challenges the idea that neighborhoods are homogeneous areas with sharp boundaries. My map highlights diversity instead of majority rule, which also means that it highlights segregation in a much more powerful way. Using dots makes it possible to ask what kinds of transitions actually exist in the city, instead of making assumptions about clean edges.

The broader lesson of my project is that we need to think more carefully about how we map social statistics, since the way we make maps influences the way we think about spatial planning and design. Right now, almost all planning is done using sharp boundaries, be they the boundaries of neighborhoods or entire countries. What would happen to our political understanding—or to our political interventions—if we made maps without solid colors? Can we imagine cartography without boundaries?

A TAXONOMY OF TRANSITIONS

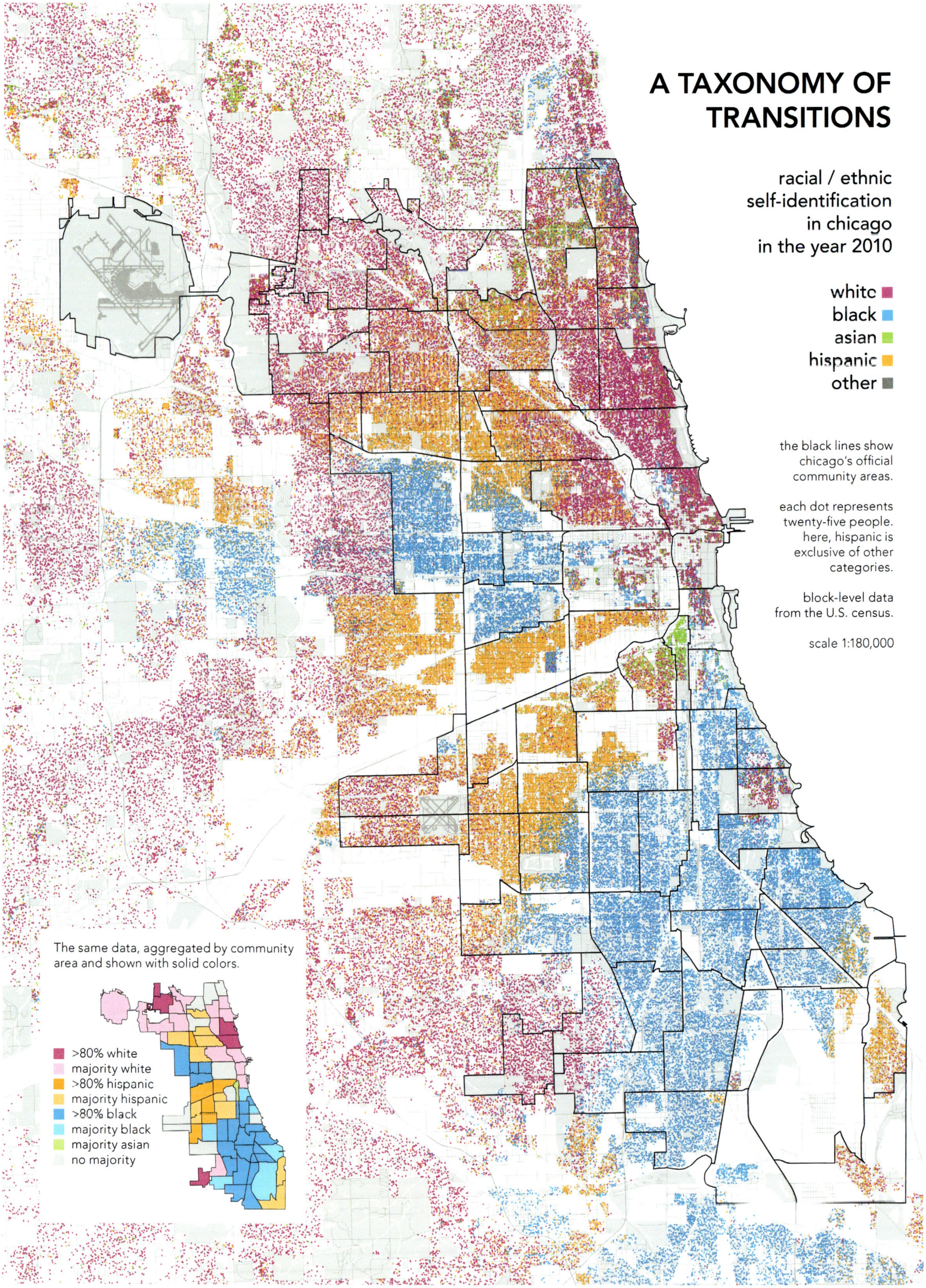

San Francisco

Racial and Ethnic Diversity in Residential Patterns

Metta Fund

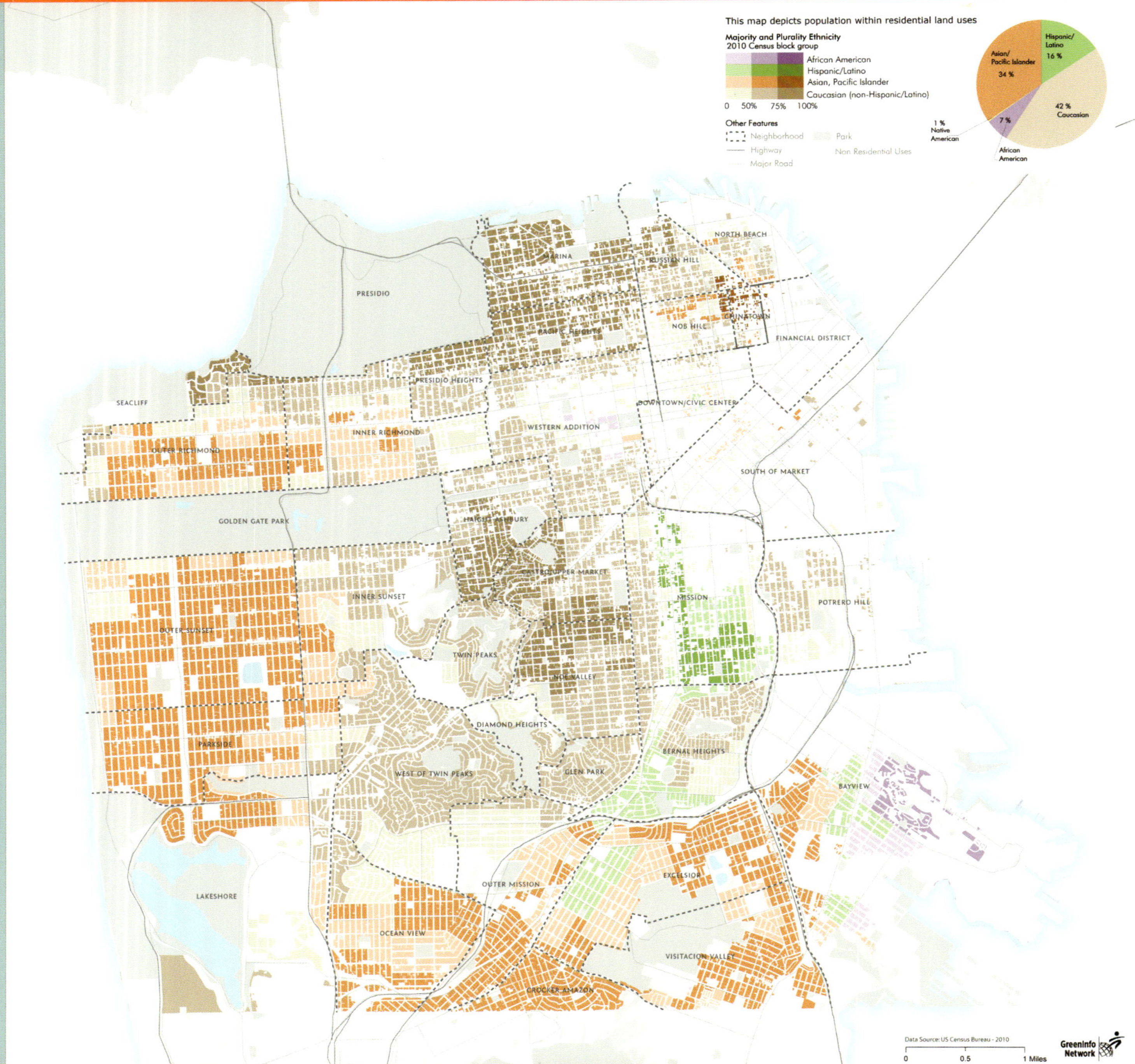

《 Racial and Ethnic Diversity in Residential Patterns

Larry Orman, Alexandra Barnish, Diana Pancholi — GreenInfo Network

This map is designed to show a *true population footprint*. It uses a creative approach to show where people *actually* live in the city of San Francisco, and what they look like. This unique style considers residential patterns and applies a "cookie cutter" to 2010 census data. The result visually removes areas of the city dominated by daytime, working traffic and defines residential neighborhoods where people actually live. The map shapes a more realistic population profile and is a block-by-block representation of San Francisco's racial and ethnic diversity.

Population information is commonly mapped over an entire landscape area. But the inclusion of streets and commercial/industrial areas can overemphasize what neighborhoods really look like. While the census must aggregate information for reporting purposes, cartographers are not limited to mapping those areas as the entire ground truth. Creative approaches are essential to understanding population patterns and improving overestimated areas.

This map's artistic perspective provides a better context for the racial and ethnic grouping of a city's residents. Featuring colors with similar saturations but opposing hues, it emphasizes patterns and diversity within the city. The green and purple colors chosen to represent minority groups (Hispanic/Latino and African American) give them a similar visual presence as the majority groups (Asian, Pacific Islander and Caucasian). Neighborhood boundaries are overlaid for context, as many people identify with San Francisco's micro-communities. A light coastal fade is used to accent the city's unique geography at the tip of a peninsula, without drawing the viewer's focus to the surrounding water.

Creative approaches are essential to understanding population patterns and improving overestimated areas.

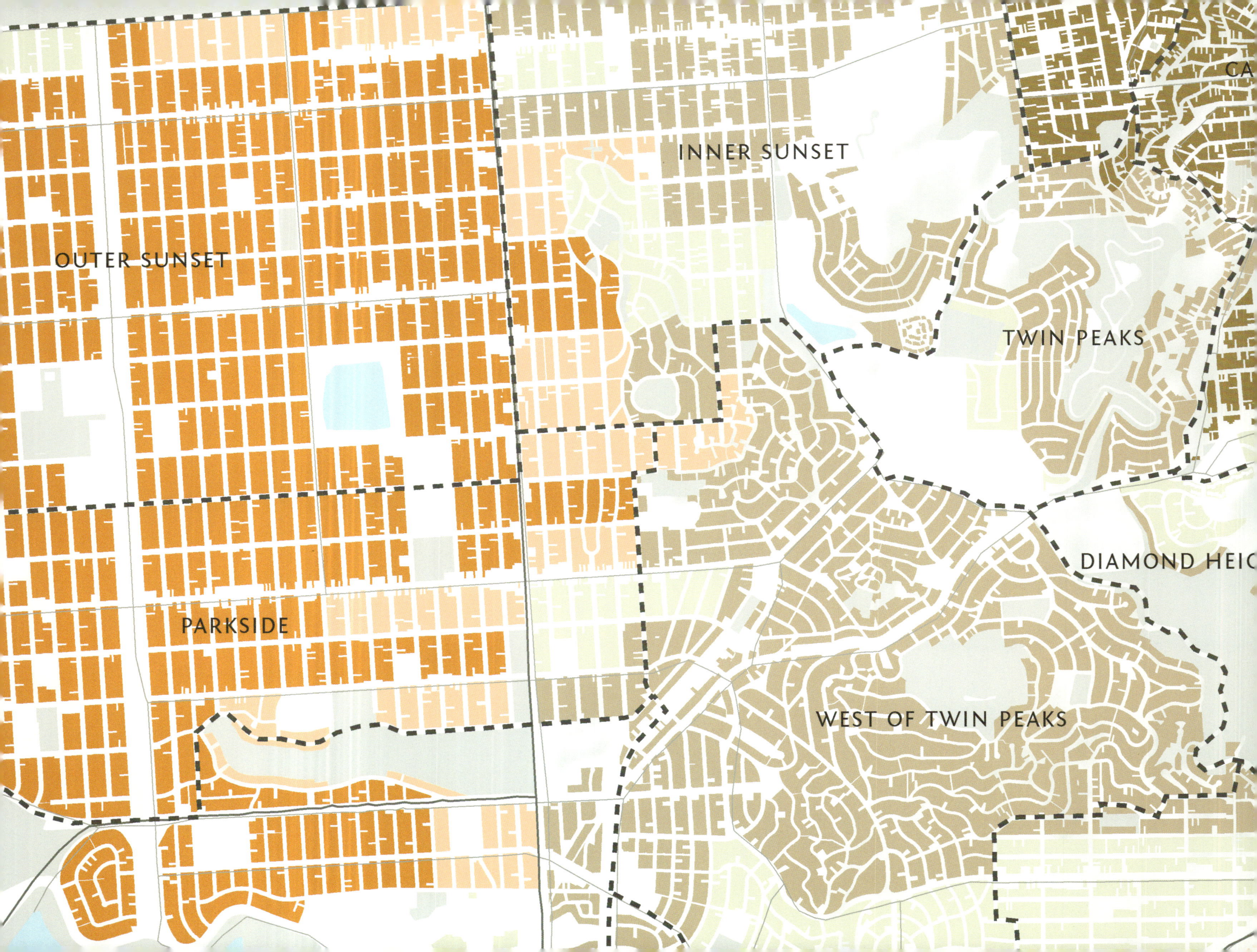

CA
INNER SUNSET
OUTER SUNSET
TWIN PEAKS
DIAMOND HEIG
PARKSIDE
WEST OF TWIN PEAKS

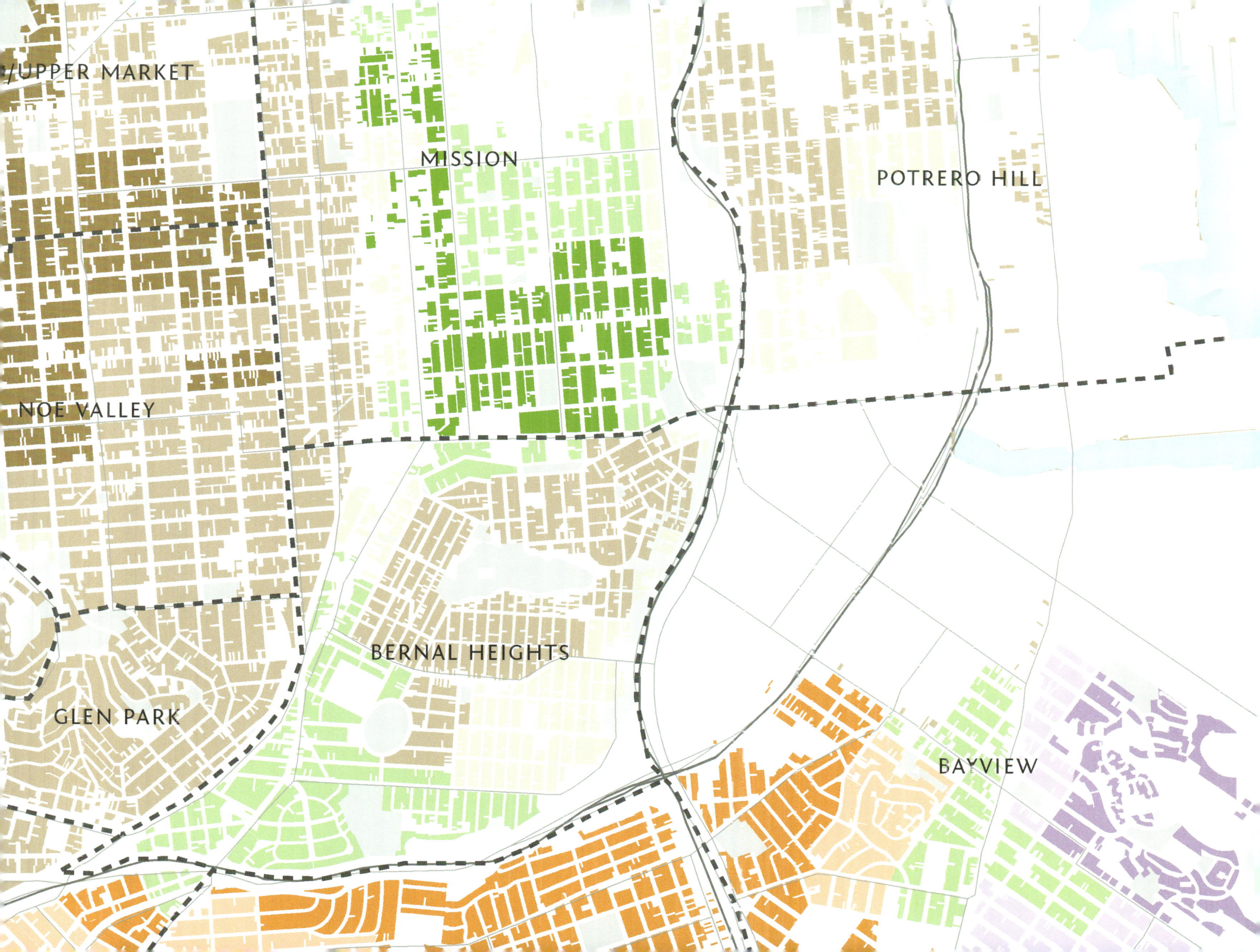

/UPPER MARKET
MISSION
POTRERO HILL
NOE VALLEY
BERNAL HEIGHTS
GLEN PARK
BAYVIEW

atlasofdesign.org